SpringerBriefs in Applied Sciences and Technology

T0185060

For further volumes:
http://www.springer.com/series/8884

Ryszard Bartnik

The Modernization Potential of Gas Turbines in the Coal-Fired Power Industry

Thermal and Economic Effectiveness

 Springer

Ryszard Bartnik
Faculty of Mechanical Engineering
Opole University of Technology
Opole
Poland

ISSN 2191-530X   ISSN 2191-5318   (electronic)
ISBN 978-1-4471-4859-3   ISBN 978-1-4471-4860-9   (eBook)
DOI 10.1007/978-1-4471-4860-9
Springer London Heidelberg New York Dordrecht

Library of Congress Control Number: 2012954066

Printed on acid-free paper

Springer is part of Springer Science+Business Media (www.springer.com)

Contents

Notations

$\dot{B}$	Stream of exergy, MW
e_{el}, e_{coal}, e_g	Specific price of electricity, coal, gas, PLN/MWh, PLN/GJ
$\dot{E}_{ch}$	Stream of chemical energy of fuel, MW
$E_{ch, A}$	Annual use of chemical energy of fuel, MWh
$E_{el, A}$	Annual production of electricity, MWh
EF	Emission factor, kg_{CO2}/MWh
h	Specific enthalpy, J/kg
H	Enthalpy, J
j	Specific capital expenditure, PLN/unit
J	Capital expenditure, PLN
k_{el}	Specific cost of electricity production, PLN/MWh
K_{cap}	Annual capital cost, PLN
K_e	Annual exploitation cost, PLN
$\dot{m}$	Stream of mass, kg/s
N	Power output, MW
NCV	Net calorific value, J/kg
p	Specific charge for pollutant emission, PLN/kg
p	Pressure, Pa
$\dot{P}$	Stream of fuel, kg/s
Q_A	Annual heat output, MWh
$\dot{Q}$	Stream of heat, J/s
s	Specific entropy, J/(kgK)
S	Entropy, J/K
$\dot{S}$	Stream of entropy, W/K
t, T	Temperature, °C, K
$z\rho + \delta_{serv}$	Annual rate of investment service and remaining fixed cost relative to capital expenditure, %/a
z	Discount rate of assets

Greek Symbols

Δ Increase

η Efficiency

Θ Exergetic temperature

ρ Annual rate of progressive depreciation, %/a

ρ Emission of pollutant per specific unit of the chemical energy
of the fuel, kg/GJ

ε_{el} Relative coefficient of power station internal load

τ Time, s

Subscripts and Superscripts

amb   Refers to cold reservoir (environment)

A   Refers to year

c   Refers to heat

C   Refers to Carnot engine

ch   Refers to chemical variables

com   Refers to combustion

dhw   Refers to domestic hot water

el   Refers to electricity

El   Refers to power plant

env   Refers to environment

ex   Refers to value before modernization

fg   Refers to exhaust gas

GT   Refers to gas turbine

h   Refers to hot reservoir

h   Refers to hot water

in   Refers to input

mod   Refers to value after modernization

n   Refers to nominal conditions

out   Refers to output

r   Refers to return water

s   Refers to steam

ST   Refers to steam turbine

u   Refers to utility heat

w   Refers to water

Chapter 1
Introduction

Abstract This chapter discusses the feasibility and the possible ways of adapting the existing coal-fired power stations to the so-called clean coal technologies based on gas turbines, including dual-fuel gas-steam combined-cycle technology with power units coupled in series and parallel systems.

Keywords Coal-fired power unit · Repowering · Gas turbine · Heat recovery steam generator · Dual-fuel combined-cycle

> For the things of this world cannot be made
> known without a knowledge of mathematics
> (Roger Bacon, 1214–1294)

This monograph undertakes a thermodynamic and economic analysis of modernization of an existing coal-fired condensing power stations to a high-efficiency, dual-fuel gas-steam system by means of their repowering with a gas turbogenerator [1, 2]. The dual-fuel, gas and coal power stations are considered as operating under clean coal technology. The modernization of the power plants by their repowering with a gas turbogenerator will, therefore, enable the rational and ecological use of coal, whose resources are considerably large in the world.

Among the possible standard gas and steam dual-fuel solutions, it is possible to identify two basic configurations (Fig. 1.1) [1, 2]:

- *in-series coupled systems* (so called Hot Windbox systems; exhaust gas from the gas turbogenerator are directed as an oxidant into the combustion chamber of the existing coal-fired boiler; this system does not include a heat recovery steam generator)
- *parallel coupled systems* (by means of a steam-water systems; the coupling involves the generation of steam in the heat recovery steam generator which is fed by exhaust gas from the gas turbogenerator into the existing collector, and (or) superheating of interstage steam in the heat recovery steam generator and (or) heating of system water in exhaust gas-water heaters dedicated for this in

R. Bartnik, *The Modernization Potential of Gas Turbines in the Coal-Fired Power Industry*, SpringerBriefs in Applied Sciences and Technology, DOI: 10.1007/978-1-4471-4860-9_1, © The Author(s) 2013

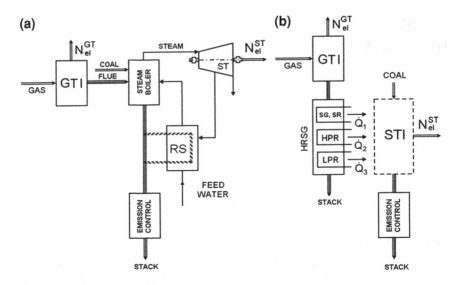

Fig. 1.1 Diagrams of dual fuel gas-steam combined-cycle: **a** in series system (Hot Windbox), **b** in parallel system. GTI—gas turbine installation, HRSG—heat recovery steam generator, ST—steam turbine, RS—regeneration system, STI—steam turbine installation, SG—steam generation, SR—steam reheating, LPR, HPR—low pressure heaters and high pressure regeneration, N_{el}^{GT}, N_{el}^{ST}—power of gas turbine and power of steam turbine

the heat recovery steam generator (thereby, the partly or totally excludes existing regenerative feed water preheaters). The heat recovery steam generator thus partly replaces the existing coal-fired boilers in the power station.

The opportunity of repowering the existing condensing power stations by means of gas turbogenerators offers an important opportunity since such modernization will result in a considerably improvement of the energy efficiency of such power stations. The thermal cycle will undergo a considerable change. Beside the current Clausius-Rankine cycle of the steam turbine the power plant will additionally apply Joule cycle which represents the operation of a gas turbogenerator engine. In addition, the repowered power plant will demonstrate a considerable, even two times increase in terms of electric power output. Additionally, the emission of pollutants into the environment will decrease as a result of lower use of coal. Moreover, in the case of a deficit of power in the transmission network, its fast increase will be possible without the necessity of building new ones. Hence, other problems connected with the social, economic, ecological, technological issues and other problems associated with the choice of a location and construction of the new power plants can be avoided.

The heat recovery steam generator forms an integral part in the high efficiency dual-fuel gas and steam systems based on a gas turbogenerator, as Joule cycle of the gas turbogenerator and Clausius-Rankine cycle of the steam turbine are coupled in it. The heat recovery steam generator, which applies low-temperature enthalpy of the exhaust gas from the gas turbogenerator for the production of the steam (its thermal parameters are imposed by the existing coal firing system) has to made to measure for each particular combined system. This is already possible under the current technology. The power output from the repowered unit is therefore dependent on the capacity of the gas turbogenerator, type of the waste heat boiler, i.e. number of pressure stages (number of evaporators), type and arrangement of the heated surfaces. An adequate selection of the structure of heat recovery steam generator leads to an increase of the power of steam turbine in the repowered unit. This value is equal to the reduction of exergy losses in the heat recovery steam generator. Therefore, the energy and economic efficiency of the repowered unit is also dependent on the type of the waste heat boiler. In the boiler the following heater installation possibilities have to be analyzed: surfaces for the generation of high-, intermediate- and low-pressure steam, surface of interstage superheater, surfaces for the high- and low-pressure regenerative feed water pre-heaters. The number of the combinations for the selection of the surfaces and their arrangement is considerable. Among them it is necessary to reject ones which could lead to hazardous overloading of the steam turbine.

This monograph presents the results of multiple alternative of calculation of thermodynamic parameters of repowering a unit with the capacity of 370 MW [2]. The analysis involves the variability of the operating conditions after it is repowered by a gas turbogenerator and single-, dual- and triple-pressure heat recovery steam generator.

Acknowledgments The author wishes to sincerely thank Dr. Anna Duczkowska-Kądziel and Dr. Zbigniew Buryn for their help during the numerical calculations.

References

1. Bartnik R (2009) Combined cycle power plants. Thermal and economic effectiveness. WNT, Warszawa
2. Bartnik R, Buryn Z (2011) Conversion of coal-fired power plants to cogeneration and combined-cycle. Thermal and economic effectiveness. Springer, London

Chapter 2
Thermodynamic Fundamentals for Production of Electric Power in Hierarchical j-Cycle Systems

Abstract This chapter presents thermodynamic fundamentals for the generation of power in the hierarchical j-cycle systems.

Keywords Thermodynamic fundamentals · Hierarchical j-cycle systems

The highest theoretical efficiency of generating mechanical energy (and, as a consequence, electric energy) in thermodynamic systems can be achieved by adopting Carnot cycle—Fig. 2.1.

This efficiency is expressed by the equation:

$$\eta_C = 1 - \frac{T_{amb}}{T_h} \tag{2.1}$$

and the power output from Carnot heat engine by the equation:

$$N_C = \eta_C \dot{Q}_d \tag{2.2}$$

where:

$\dot{Q}_d$ stream of the driving heat,
T_{amb} absolute temperature of the cold reservoir (environment),
T_h absolute temperature of the hot reservoir.

The power N_C is equivalent with the exergy stream $\dot{B}$ of stream of heat $\dot{Q}_d$ transferred from the source with the temperature of $T_h = $ const, $N_C \equiv \dot{B}_{\dot{Q}_d}$.

If a power plant were to realize the Carnot cycle (which is technically impossible), for the temperatures $T_h = T_{com} = 1600$ K and $T_{amb} = 300$ K its efficiency, under the assumption of a lack of losses during the conversion of mechanical energy into electric energy would be $\eta_C = 81$ % (T_{com} denotes the temperature of the combustion of coal in the boiler). Concurrently, the gross

R. Bartnik, *The Modernization Potential of Gas Turbines in the Coal-Fired Power Industry*, SpringerBriefs in Applied Sciences and Technology, DOI: 10.1007/978-1-4471-4860-9_2, © The Author(s) 2013

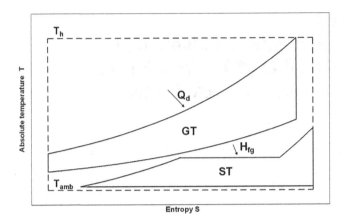

Fig. 2.1 Comparative cycle (theoretical) of a single-fuel gas–steam system (GT—Joule's cycle of the gas turbine, ST—Clausius-Rankine cycle of the steam turbine, Q_d—driving heat transferred into GT, H_{fg}—enthalpy of flue gas exiting from the gas turbine transferred to ST through heat recovery steam generator; *dashed line* marks Carnot cycle for the extreme temperatures T_{amb} and T_h)

efficiency of a steam based Clausius-Rankine cycle realized in a power station is smaller by around 50 %. For instance, for a 370 MW power unit operating under subcritical parameters this efficiency is equal to mere 41 %.

From the equation in (2.1), a conclusion can be made that the same stream of heat $\dot{Q}_d$ transferred from the source with the temperature of $T_h = $ const can be transformed into mechanical power to the greater degree the higher the value of temperature T_h. The power (stream of exergy) losses as a result of lowering the temperature from T_{h1} to T_{h2} ($T_{h1} > T_{h2}$) is equal to:

$$\Delta N_C = \delta \dot{B} = \dot{Q}_d \left(1 - \frac{T_{amb}}{T_{h1}} \right) - \dot{Q}_d \left(1 - \frac{T_{amb}}{T_{h2}} \right) = T_{amb} \dot{Q}_d \frac{T_{h1} - T_{h2}}{T_{h1} T_{h2}}. \quad (2.3)$$

The value on the right hand side of Eq. (2.3) concurrently denotes the loss of exergy stream in the irreversible heat transfer between the two sources with the temperatures of $T_{h1} = $ const and $T_{h2} = $ const. This loss can be expressed also in terms of the increase of their entropy streams (compare Eq. 2.8).

By analogy to formula (2.1), the energy efficiency of any cycle can be expressed by the equation [1]:

$$\eta_t = 1 - \frac{\overline{T}_{out}}{\overline{T}_{in}} \quad (2.4)$$

where the temperature T_h of the isotherm of the Carnot cycle that is equal to the temperature of the hot reservoir can be replaced by the entropy averaged temperature $\overline{T}_{in}$ during the transfer of heat into a medium in an arbitrarily considered cycle (Eq. 2.5), and the temperature T_{amb} of the isotherm of the Carnot cycle of the cold

Fig. 2.2 Thermodynamic cycle

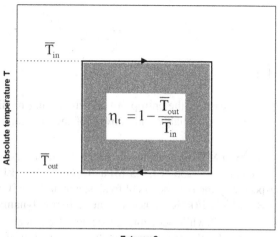

reservoir, i.e. the environment, is replaced by the entropy averaged temperature $\overline{T}_{out}$ during the extraction of heat from an arbitrary cycle—Fig. 2.2.

The introduction of mean thermodynamic temperatures $\overline{T}_{in}$ and $\overline{T}_{out}$ for the subsequent input and output of heat from a system (which are calculated for the actual temperatures and pressures at the beginning and output from these processes, i.e. for irreversible processes), makes it possible to present any cycle in a *temperature-entropy* co-ordinate system in the form of a rectangular shape (Fig. 2.2), regardless of the nature of the processes of the physical changes occurring during them, including those in which actual effective work is exerted, whether reversible or not. The exergy losses as a result of the friction during these conversions only have to be involved in the mean values of the entropy changes at temperatures of the heat input $\overline{T}_{in}$ and output $\overline{T}_{out}$ from a system only in the case if predefined correction measures are adopted. The quotient of the averaged temperatures during these conversions has to apparently equal to the quotient of the heat output Q_{out} and input Q_{in} into a given cycle [1].

From the relation in (2.4) it stems that the generation of electricity in the cycles of thermal power plants should be undertaken for a technically maximum temperature $\overline{T}_{in}$ of the circulating medium during the input of stored heat, i.e. heat from an external source and for the lowest temperature $\overline{T}_{out}$ of the medium output of heat from the cycle in a power plant.

Clausius-Rankine cycle is followed in a coal-fired power plants (Fig. 2.1). From the thermodynamic perspective, its fundamental drawback is associated with the low mean thermodynamic temperature $\overline{T}_{in}$ of the circulating media—water and steam (also called the entropy averaged temperature) during isobaric process of the heat Q_{in} transfer into this cycle in the boiler (compare Eqs. 2.6, 2.7):

$$\overline{T}_{in} = \frac{Q_{in}}{\Delta S} = \frac{\int_{s_w}^{s_s} T_{in}(s)ds}{s_s - s_w} = \frac{h_s - h_w}{s_s - s_w} \tag{2.5}$$

where:

ΔS increase of the entropy of the circulating medium,

h, s specific enthalpy and entropy of the circulating medium.

In a 370 MW power unit, the thermal parameters of the water fed into the boiler are equal to: 255 °C/23.5 MPa ($h_w = 1110.8$ kJ/kg, $s_w = 2.7947$ kJ/(kgK)), respectively the parameters of fresh steam are 535 °C/18 MPa ($h_s = 3373.2$ kJ/kg, $s_s = 6.3537$ kJ/(kgK); hence, the mean thermodynamic temperature is equal to only $\overline{T}_{in} = 635$ K (while accounting for inter-stage steam superheating $\overline{T}_{in} = 640.7$ K). Concurrently, the temperature of the combustion of coal in the boiler is equal to around $T_{com} = 1600$ K. Therefore, the temperature difference $T_{com} - \overline{T}_{in} \cong 1000\,K$ is considerable, which along with the low temperature $\overline{T}_{in}$ results in small efficiency of generating electricity in a power unit (from Eqs. 2.1–2.4) it stems that it is equal to mere 41 % gross (which is 37 % net).

The power losses expressed by the Eq. (2.3) takes place in the steam boiler, while $\dot{Q}_d = \dot{E}_{ch}^{coal}$ ($\dot{E}_{ch}^{coal}$ denotes the stream of chemical energy of the coal) is equal to the product of the stream of coal combustion in the boiler $\dot{P}$ and its net calorific value NCV, $\dot{Q}_d = \dot{E}_{ch}^{coal} = \dot{P}(NCV)$, and temperatures T_{h1} and T_{h2} are equal to: $T_{h1} = T_{com}$ and $T_{h2} = \overline{T}_{in}$. In terms of numbers, the loss of power is equal to 30 % of the driving heat $\dot{Q}_d$: $T_{amb}(T_{com} - \overline{T}_{in})/(T_{com}\overline{T}_{in}) = 300(1600 - 636)/(1600 \times 636) \cong 30\%$. Thus, despite its high energy efficiency, reaching 94 %, the steam boiler forms the major source of the low efficiency of generating electric power in steam power plants operating in the Clausius-Rankine cycle.

However, the considerable advantage of the Clausius-Rankine cycle is the low temperature $\overline{T}_{out}$ in it. The condensation isotherm in it nearly overlaps with the isotherm of the ambient temperature in the Carnot cycle, $\overline{T}_{out} \cong T_{amb}$ (Fig. 2.1).

The use of the higher range of temperature, even starting from the temperature of gas combustion $t_{com} = 1500$ °C is taken advantage of in gas turbogenerators. The production of electric power in it occurs by direct expansion of exhaust gas from the temperature and pressure in the combustion chamber to the ambient pressure. Hence, a coupling of the steam system with the gas system, whose advantage involves a considerably higher temperature $\overline{T}_{in}$ compared with the steam boiler (the disadvantage of the gas system involves also the high temperature $\overline{T}_{out}$ of the circulating medium during the extraction of heat from it), results in the use of the advantages of the two cycles while avoiding their drawbacks. As a result, the efficiency of producing electricity in power plants adapted to a gas-steam system considerably increases. The device which couples the two cycles is the heat recovery steam generator—Fig. 1.1b. The steam produced in it has identical thermal parameters as the steam from the coal-fired boiler. The total stream of fresh steam from the heat recovery steam generator and coal-fired boiler is equal to the stream of steam prior to when the system was not repowered. The production of steam in the heat recovery steam generator applies the stream of low-temperature enthalpy of the flue gas H_{fg} from the gas turbine. Thus, its enthalpy partly replaces the use of coal in the existing coal-fired system, due to which the use of coal is limited. As a result, the loss of the unused higher range of temperatures $T_{com} - \overline{T}_{in} \cong 1000\,K$ is reduced. As an additional consequence, the efficiency of the generation of electricity in dual-fuel gas-steam systems is improved. This efficiency increases along with the increase of the capacity of the gas turbine and it can reach as much as by 10 %.

The highest efficiency (Eq. 2.19), even as much as 60 %, is possible in single-fuel gas-steam systems (Figs. 2.1, 2.4) [1], where the coal-fired boiler is excluded and, thus, the phenomenon of unused higher range of the temperature of the flue gas is avoided. The total driving heat Q_d from the combustion of gas (or liquid fuel; a very attractive concept in terms of energy and economic efficiency involves direct coal combustion in a gas turbine) is input into a gas turbine operating under the Joule's cycle (Fig. 2.1). The steam-based section still operates in the Clausius-Rankine cycle, but the driving heat for the production of steam originates only from the low-temperature enthalpy of the flue gas H_{fg} extracted from the gas turbine. As a result, the loss of the unused higher range of temperatures is avoided while the efficiency of the production of electricity increases in comparison to the system solely based on steam. Such an increase of efficiency can be explained in a form of a chart. As one can see in Fig. 2.1, the Carnot cycle is supplemented by the Joule's cycle, as a result of which there is a considerable reduction of the surface areas of the conversion phenomena in the Clausius-Rankine cycle and Carnot cycle.

In a general case, the number of circulating media can be arbitrarily large. A hierarchical j-cycle system is presented in Fig. 2.3. An increase of the number of media with various temperatures of the operating range makes it possible to apply in a system higher range of the temperature increase between the upper and lower heat sources (environment). Thereby, exergy losses in the system are reduced and the production of electricity increases. The disadvantage of such a solution includes an increase of investment required to start the system.

Generally, the loss of exergy stream $\delta \dot{B} = T_{amb} \left(\sum_k \Delta \dot{S}_{med} + \sum_l \Delta \dot{S}_{so} \right)$ in a hierarchical "j-cycle" system comes as a consequence of mere increase of entropy streams of external heat sources $\sum_l \Delta \dot{S}_{so}$ which are in contact with it [1] (in practice we usually have to do with two sources, $l = 2$). The substitution of actual open cycle processes by closed-loop system, which normally facilitates the thermodynamic analysis of such processes, leads to a lack of consideration of the media input and output from the system; hence, the increase of entropy streams is equal to zero, $\sum_k \Delta \dot{S}_{med} = 0$. Hence, an increase of the entropy of the bodies which participate in the phenomenon is expressed only in terms of the increase of the entropy of heat

Fig. 2.3 Diagram of hierarchical j-cycle heat engine

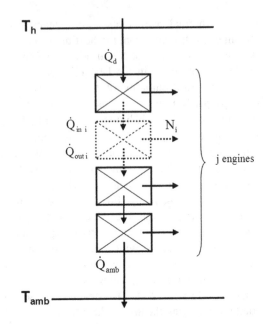

sources. This increase can then be expressed in terms of the total of entropy increases in irreversible heat flow between the sources and cycles as well as between the cycles.

One can note in this place that the increase of entropy of the external source of heat with the temperature $T_{so} = $ const which delivers heat Q into the system can be derived from the definition of entropy

$$\Delta S_{so} = -\int \frac{dQ}{T_{so}} = -\frac{Q}{T_{so}}. \tag{2.6}$$

The minus sign in Eq. (2.6) denotes that the positive heat Q was extracted from the source. For the source which pulls heat Q from the system, it is only necessary in Eq. (2.6) to change the sign

$$\Delta S_{so} = +\int \frac{dQ}{T_{so}} = +\frac{Q}{T_{so}}. \tag{2.7}$$

By applying the Eqs. (2.6) and (2.7) the loss of exergy stream $\delta \dot{B}$ in a closed system with two heat sources with the temperatures of T_h and T_{amb} (Fig. 2.3) can be expressed by the equation [1] (compare Eq. 2.3):

$$\delta \dot{B} = T_{amb} \sum_{2} \Delta \dot{S}_{so} = T_{amb} \left(\frac{\dot{Q}_{amb}}{T_{amb}} - \frac{\dot{Q}_d}{T_h} \right) = \sum_{i=1}^{j+1} \delta \dot{B}_i$$

$$= T_{amb} \sum_{i=1}^{j+1} \dot{Q}_{in\,i} \frac{\overline{T}_{out\,i-1} - \overline{T}_{in\,i}}{\overline{T}_{out\,i-1}\overline{T}_{in\,i}}, \tag{2.8}$$

and the capacity of the system by the equation:

$$N = N_C - \delta\dot{B} = \dot{Q}_d \frac{T_h - T_{amb}}{T_h} - T_{amb}\left(\frac{\dot{Q}_{amb}}{T_{amb}} - \frac{\dot{Q}_d}{T_h}\right)$$

$$= \sum_{i=1}^{j} N_i = \sum_{i=1}^{j} (\dot{Q}_{in\,i} - \dot{Q}_{out\,i}) = \sum_{i=1}^{j} \dot{Q}_{in\,i} \frac{\overline{T}_{in\,i} - \overline{T}_{out\,i}}{\overline{T}_{in\,i}} \qquad (2.9)$$

$$= \dot{Q}_d \frac{T_h - T_{amb}}{T_h} - T_{amb} \sum_{i=1}^{j+1} \dot{Q}_{in\,i} \frac{\overline{T}_{out\,i-1} - \overline{T}_{in\,i}}{\overline{T}_{out\,i-1}\overline{T}_{in\,i}}$$

where:

j	number of circulating media (engines),
N_c, N_i	capacity of a theoretical Carnot engine and actual engines,
$\dot{Q}_{in\,i}, \dot{Q}_{out\,i}$	heat of stream input into and output from an i-th cycle (engine), while $\dot{Q}_{out\,i} = \dot{Q}_{in\,i+1}$ and $\dot{Q}_{in\,1} \equiv \dot{Q}_d$, $\dot{Q}_{in\,j+1} \equiv \dot{Q}_{amb}$,
$\dot{Q}_{amb}, \dot{Q}_d$	stream of heat transmitted from the system into the environment and delivered from the upper source of heat,
$\overline{T}_{in\,i}, \overline{T}_{out\,i}$	mean thermodynamic temperature of the absorbing medium, and giving off heat in an i-th cycle (engine), while $\overline{T}_{in\,j+1} \equiv T_{amb}, \overline{T}_{out\,0} \equiv T_h$,
T_h	absolute temperature of the upper source of heat.

The value of $(\overline{T}_{in\,i} - \overline{T}_{out\,i})/\overline{T}_{in\,i}$ in the final term of the Eq. (2.9) represents the energy efficiency of an i-th ($i = 1 \div j$) engine operating between entropy averaged temperatures in actual processes of heat input and output $\overline{T}_{in\,i}$, $\overline{T}_{out\,i}$ from a system (compare Eq. 2.4, Fig. 2.2):

$$\eta_i = 1 - \frac{\overline{T}_{out\,i}}{\overline{T}_{in\,i}} \qquad (2.10)$$

The stream of heat output from an i-th cycle (engine) by means of entropy averaged temperatures can only be expressed by streams of heat $\dot{Q}_{in\,1} \equiv \dot{Q}_d$ delivered into the system from the source with the temperature $\overline{T}_{out\,0} \equiv T_h$

$$\dot{Q}_{out\,i} = \dot{Q}_{in\,i+1} = \dot{Q}_d \prod_{n=1}^{i} \frac{\overline{T}_{out\,n}}{\overline{T}_{in\,n}}. \qquad (2.11)$$

From the Eq. (2.11) we obtain the relation for the heat output from the system into the environment

$$\dot{Q}_{amb} = \dot{Q}_d \prod_{i=1}^{j} \frac{\overline{T}_{out\,i}}{\overline{T}_{in\,i}}. \qquad (2.12)$$

Fig. 2.4 Diagram of hierarchical, 2-cycle, gas–steam heat engine

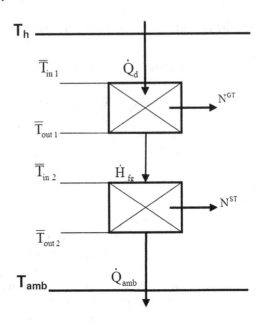

By applying the relation (2.11) to the equation in (2.9) the value of the total power of the system can be defined as:

$$N = \sum_{i=1}^{j} N_i = \dot{Q}_d \left(1 - \prod_{i=1}^{j} \frac{\overline{T}_{out\,i}}{\overline{T}_{in\,i}} \right). \tag{2.13}$$

The value in the brackets on the right hand side of Eq. (2.13) denotes the energy efficiency η_{1-j} of generating power in a system with j-cycles expressed by entropy averaged temperatures:

$$\eta_{1-j} = 1 - \prod_{i=1}^{j} \frac{\overline{T}_{out\,i}}{\overline{T}_{in\,i}}. \tag{2.14}$$

For instance, for a two-cycle system this efficiency, by additionally using relation (2.10), can be expressed by the equation:

$$\eta_{1-2} = 1 - \frac{\overline{T}_{out\,1}}{\overline{T}_{in\,1}} \frac{\overline{T}_{out\,2}}{\overline{T}_{in\,2}} = \eta_1 + \eta_2 - \eta_1\eta_2. \tag{2.15}$$

The final form of the Eq. (2.8) which distinguishes the location of the origin of exergy losses in the system makes it possible to find ways of its thermodynamic improvement. It indicates the places of greatest exergy losses which determine its low effectiveness, and therefore, indicates the places in which it could be improved. It also indicates the entropy averaged temperatures and direction of altering its values so as to improve the thermodynamic perfection of the system.

In addition, the presented equation makes the quantitative analysis of the reasons which increase this perfection. Furthermore, the analysis of the variability of parameters of the preceding cycles is possible resulting in a change to exergy losses in the subsequent phases, thereby affecting exergy losses in the whole system. Thus, it is possible to undertake the justification for these changes.

The final form of the Eq. (2.8) makes it possible to analyze the effect of energy efficiency of the particular engines (entropy averaged temperatures of circulating media) on the total energy efficiency of the system.

For the case of a two-cycle system ($j = 2$), (Figs. 2.1, 2.4), the exergy losses as a result of irreversible heat flow between the sources and circulating media (Eq. 2.8), the total power of the system (Eq. 2.9) and the energy efficiency can be expressed by the subsequent equations:

- exergy losses

$$\delta \dot{B}^{G-S} = T_{amb} \left(\dot{Q}_d \frac{T_h - \overline{T}_{in\,1}}{T_h \overline{T}_{in\,1}} + \dot{H}_{fg} \frac{\overline{T}_{out\,1} - \overline{T}_{in\,2}}{\overline{T}_{out\,1} \overline{T}_{in\,2}} + \dot{Q}_{amb} \frac{\overline{T}_{out\,2} - T_{amb}}{\overline{T}_{out\,2} T_{amb}} \right) \quad (2.16)$$

- power of the system

$$N^{G-S} = N^{GT} + N^{ST} = \dot{Q}_d \frac{\overline{T}_{in\,1} - \overline{T}_{out\,1}}{\overline{T}_{in\,1}} + \dot{H}_{fg} \frac{\overline{T}_{in\,2} - \overline{T}_{out\,2}}{\overline{T}_{in\,2}} = \dot{Q}_d \eta_{GT} + \dot{H}_{fg} \eta_{ST} \quad (2.17)$$

or by applying the final form of the Eq. (2.9)

$$N^{G-S} = N_C - \delta \dot{B}^{G-S}$$

$$= \dot{Q}_d \frac{T_h - T_{amb}}{T_h} - T_{amb} \left(\dot{Q}_d \frac{T_h - \overline{T}_{in\,1}}{T_h \overline{T}_{in\,1}} + \dot{H}_{fg} \frac{\overline{T}_{out\,1} - \overline{T}_{in\,2}}{\overline{T}_{out\,1} \overline{T}_{in\,2}} + \dot{Q}_{amb} \frac{\overline{T}_{out\,2} - T_{amb}}{\overline{T}_{out\,2} T_{amb}} \right) \quad (2.18)$$

- energy efficiency (compare Eqs. 2.10, 2.15)

$$\eta_{G-S} = \frac{N^{GT} + N^{ST}}{\dot{E}_{ch}^{gas}} = \eta_{GT} + \eta_{ST} - \eta_{GT} \eta_{ST} \quad (2.19)$$

where:

$\dot{E}_{ch}^{gas} \equiv \dot{Q}_d$ stream of chemical energy of the gas combustion in the gas engine,

N^{GT}, N^{ST} power of the gas and steam engines,

η_{GT}, η_{ST} energy efficiency of the gas and steam engines,

$\overline{T}_{in\,1}, \overline{T}_{out\,2}$ subsequent mean thermodynamic temperature in the combustion chamber of the gas turbine and temperature of steam saturation in the condenser of the steam turbine,

$\overline{T}_{out\,1}, \overline{T}_{in\,2}$ subsequent mean thermodynamic temperature of flue gas and steam in the heat recovery steam generator.

Only the value of temperature $\overline{T}_{in\,2}$ can be determined by the designer by means of altering the number of the heating surfaces, as well as their design and sizes, their location as well by means of adopting temperature intervals, i.e. differences between the temperature of flue gas and water and steam in the process heat exchange in the waste boiler. Concurrently, the values of temperatures $\overline{T}_{in\,1}, \overline{T}_{out\,1}$ are relative only to the type of turbogenerator used in a given system and there is no way of affecting them, as well as no effect that can be possibly made to temperature $\overline{T}_{out\,2}$, which is relative to ambient temperature.

The expression (compare Eq. 2.3)

$$T_{amb}\dot{H}_{fg}\frac{\overline{T}_{out\,1} - \overline{T}_{in\,2}}{\overline{T}_{out\,1}\overline{T}_{in\,2}} \tag{2.20}$$

in Eq. (2.18) denotes the loss of exergy stream in the heat recovery steam generator $\delta\dot{B}_{HRSG}$ (compare 6.2). The higher the temperature $\overline{T}_{in\,2}$, which can be determined by the designer, the smaller the losses and, thereby, the greater the electrical capacity of the steam turbogenerator (Eq. 2.17)

$$N^{ST} = \dot{H}_{fg}\frac{\overline{T}_{in\,2} - \overline{T}_{out\,2}}{\overline{T}_{in\,2}} = \dot{H}_{fg}\eta_{ST}. \tag{2.21}$$

To summarize, the structure of the heat recovery steam generator should be adopted in a way that ensures that the equation in (2.20) assumes the lowest possible value. The decrease in the value of (2.20), which means the reduction of exergy losses in a repowered unit, the greater the increase in the production of electricity in a unit. However, investment required for the power unit is greater, which can result in the limitation of the economic efficiency of the operation of the repowered unit (Eq. 6.3). Therefore, there is a technical and economic optimum, which has to be sought.

Reference

1. Bartnik R (2009) Combined cycle power plants. Thermal and economic effectiveness. WNT, Warszawa

Chapter 3
In-Series or Parallel System?

Abstract This chapter discusses the advantages and drawbacks of modernizing the existing coal-fired condensing power stations to a high-efficiency, dual-fuel gas-steam combined cycle by means of their repowering with a gas turbogenerator in series and parallel systems. Energy balance equations are also presented for them in the chapter.

Keywords Dual-fuel combined cycle · Energy balance · In-series system · Parallel system

The weakest link in terms of electricity generation in a power plant is the coal-fired steam boiler, despite its high energy efficiency. It is so since the boiler forms the exchange surface between the flue gas and water (which form the intermediate energy carrier in the electricity generation chain in a power plant), which generates the highest exergy losses (electrical capacity) in a power plant as a result of irreversible flow of heat. The reason for such high losses is associated with the around 1000 °C difference between the temperature of coal combustion in the boiler and the temperature of the produced steam. The energy effectiveness of repowering a power plant will be therefore the higher the greater the degree in which the exergy losses are achieved in the boiler, i.e. the lower the combustion of coal in the boiler. The thermodynamic criterion for the search of an optimum solution of repowering a power plant with a steam turbine should consist in the minimization of the exergy losses in the existing coal-fired boiler (in the in-series system, heat recovery steam generator is excluded) while accounting for such limitations, as the technically admissible reduction in the load of a coal-fired boiler and maximum overloading of a steam turbine and electric generator coupled with it. In practice, the technical minimum of a boiler is equal to 45–50 % of the nominal loading, while the overloading of the electric generator can be as much as 10 %.

The selection of a gas turbine in a Hot Windbox system to match a specific steam boiler involves the adaptation of the oxygen stream carried in the flue gases from the turbine to the technological requirements of the boiler. The mass fraction of oxygen in the flue gas ranges from $g_{O2} = 13$ to 16 % (in the air $g_{O2} = 23$ %; such a large oxygen concentration in the flue gas comes from the large of excess air coefficient λ^{GT} in the combustion chamber of the gas turbine due to the limited thermal strength of the material used in turbine blades $-\lambda^{GT} = 2, 5$–4). While the effort is aimed at the complete replacement of the combustion air with exhaust gas from the turbine (electricity output from turbogenerator is then equal to the maximum of power which is justified by thermodynamic considerations), the stream of such flue gas should be greater than the stream of air by 44 to 77 % [1].

The in-series system (Hot Windbox) (Chap. 1, Fig. 1.1a; Fig. 3.1) requires considerable adaptation of the existing coal-fired boiler due to high temperature of flue gas from the gas turbine fed coal dust burners into the combustion chamber of boiler and considerably larger mass stream of the gas in comparison to the equivalent air stream needed for coal combustion. The associated increase of the velocity of passing flue in spite of a reduced use of coal leads to erosion hazard in the heated surfaces. Hence, in the Hot Windbox it would be suitable to adept a gas turbine with a lower capacity and the deficiency of oxygen could be supplemented by the atmospheric air by means of the existing blowers (another solution could be associated with the reduction of the loading of the boiler and decrease of steam mass generated in it). This would, however, result in a smaller improvement of energy efficiency in the repowered unit. Along with the change in the velocity of flue gas, energy balances for the particular heated surfaces are altered as well. For example, it is to remove an existing air preheater, and the coal-fired boiler has to undergo deep engineering changes. Its supporting structure has to accommodate more heating surfaces. The investment in the adaptation of the boiler would be considerably larger than the comparable cost of a heat recovery steam generator. In practice, there is additionally a lack of void space for the installation of a gas turbogenerator with the air inlet system and exhaust gas duct near the boiler. In addition, Hot Windbox system requires a long, several months of downtime needed for its construction.

Such problems are not encountered in parallel coupling, as the latter offers the possibility of free choice of the capacity of a gas turbogenerator and leads to greater possibility of using enthalpy of the flue gas [1]. Furthermore, for the parallel coupling (Chap. 1, Fig. 1.1b; Fig. 3.2) there are more possibilities of reducing the use of coal in the steam boiler than in the in-series system; hence, exergy losses in the system are smaller. The needs for repowering of the steam-water system of the existing coal-fired section of the power plant are also smaller, thus, lower investment is required.

The necessary investment associated with the repowering will be associated only with the newly developed gas system as well as its coupling with the existing one. The construction of a gas system takes place when the existing coal-fired boiler is in operation. Then, there are no economic losses associated with its downtime period. In addition, the connection of the gas system with the coal-based facility will only take several days. In conclusion, the parallel system is a more efficient way of repowering a power plant in terms of energy efficiency and in economic calculation. This is the reason why a mathematical model has been developed for such a system

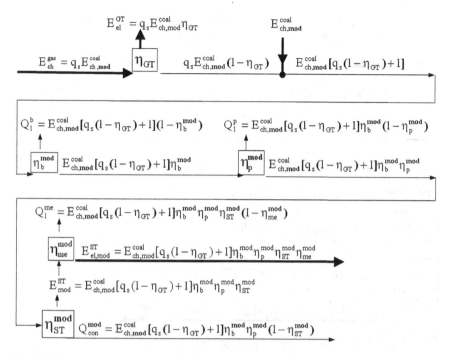

Fig. 3.1 Energy balance of a power plant repowered by a gas turbine in an in-series system

with 370 MW power unit [2]. One can note that the electric capacity of a power plant subsequent to such repowering can be even doubled.

Below is a presentation of energy balance equations for the repowered power plant in a series and parallel system:

- energy balance of a power plant repowered by a gas turbine in an in-series system (Hot Windbox; Chap. 1, Fig. 1.1a)
- energy balance of a power plant repowered by a gas turbine in a parallel system (Chap. 1, Fig. 1.1b):

where:

E_{el}^{ST}	gross electricity generation in the steam turbogenerator,
E_{el}^{GT}	gross electricity generation in the gas turbogenerator,
E_{ch}^{gas}	chemical energy of the gas combustion in the gas turbine,
$E_{ch,mod}^{coal}$	chemical energy of the coal combustion in the existing boiler in the repowered power plant,
q_p	share of the chemical energy of the gas in the parallel system in the chemical energy of coal combustion in the repowered power plant, $q_p \in (0; 1.4)$ [1] (Chap. 5, Fig. 5.3),

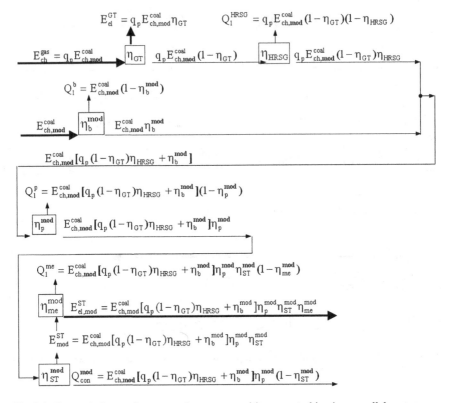

Fig. 3.2 Energy balance of a power plant repowered by a gas turbine in a parallel system

q_s share of the chemical energy of gas in the in-series system in the chemical energy of coal combustion in the repowered power plant, $q_s \in (0; 0.6)$ [1],

Q_{con} heat of condensing steam in the condenser of the steam turbine,

Q_l heat loss

η_b gross efficiency of the boiler (if E_{el}^{ST} is the net production, η_b has to be the net efficiency),

η_{HRSG} gross efficiency of the heat recovery steam generator,

η_p energy efficiency of the collector system which feeds steam into the turbine,

$\eta_{ST} = \eta_{CR}\eta_i$ energy efficiency of the steam turbine (product of the energy efficiency of Clausius-Rankine cycle and internal efficiency of the steam turbine),

$\eta_{me} = \eta_m\eta_G$ electromechanical efficiency of the turbogenerator (product of mechanical efficiency of the steam turbine and total efficiency of the electric generator),

η_{GT} gross energy efficiency of the gas turbine.

As noted above, the efficiency of a dual-fuel gas-steam fuel due to exergy losses (Chap. 2, Eq. 2.3) associated with the upper range of temperatures of flue gas in the coal-fired boiler, despite the reduced use of coal in it is lower than it is for the case of single-fuel gas-steam system. In a single-fuel system these losses are completely excluded as the production of steam occurs exclusively in the heat recovery steam generator by means of low-temperature enthalpy of exhaust gas from the gas turbine for very small differences between the temperature of flue gas and temperature of the heated medium in the range of several degrees.

References

1. Bartnik R (2009) Combined cycle power plants. Thermal and economic effectiveness. WNT, Warszawa
2. Bartnik R, Buryn Z (2011) Conversion of coal-fired power plants to cogeneration and combined-cycle. Thermal and economic effectiveness. Springer, London

Chapter 4
Energy Efficiency of Repowering a Power Unit by Installing a Gas Turbogenerator in a Parallel System

Abstract This chapter presents equations of energy efficiency for the conditions of repowering a power unit by installing a gas turbogenerator in a parallel system.

Keywords Energy efficiency · Incremental efficiency · Apparent efficiency

The repowering of a power unit by a gas turbogenerator and heat recovery steam generator considerably increases the efficiency of electricity generation as a result of adaptation of the thermal cycle realized in it. Beside the former Clausius-Rankine cycle of the steam turbine the power unit will benefit from the coupling which realizes Joule's cycle of the gas turbine. The efficiency of electricity production in the repowered unit will be expressed by the equation [1, 2]:

$$\eta_{Eel}^{p} = \frac{N_{el}^{El} + \Delta N_{el}^{ST} + N_{el}^{GT}}{\dot{E}_{ch}^{coal} + \dot{E}_{ch}^{gas}} = \frac{N_{el}^{ST} + N_{el}^{GT}}{\dot{E}_{ch}^{coal} + \dot{E}_{ch}^{gas}} \qquad (4.1)$$

where:

$\dot{E}_{ch}^{coal}, \dot{E}_{ch}^{gas}$	stream of chemical energy of coal and gas combustion in the unit after repowering of the unit,
N_{el}^{El}	capacity of the steam turbogenerator prior to repowering the unit by a gas turbine,
ΔN_{el}^{ST}	increase in the capacity of the steam turbogenerator after repowering a power unit by a gas turbine,
$N_{el}^{ST} = N_{el}^{El} + \Delta N_{el}^{TP}$	capacity of the steam turbogenerator after repowering a power unit by a gas turbine,
N_{el}^{GT}	capacity of the gas turbogenerator.

It is also possible to determine the incremental efficiency of electricity production in a repowered power unit. This efficiency is the equivalent to the efficiency of

R. Bartnik, *The Modernization Potential of Gas Turbines in the Coal-Fired Power Industry*, SpringerBriefs in Applied Sciences and Technology, DOI: 10.1007/978-1-4471-4860-9_4, © The Author(s) 2013

electricity generation (Chap. 2, Eq. 2.19) in the currently most effective thermodynamically, classical in-series gas-steam system, single fuel combined cycle
(Chap. 2, Figs. 2.1, 2.4):

$$\eta_\Delta = \frac{N_{el}^{GT} + \Delta N_{el}^{ST}}{\dot{E}_{ch}^{gas}}. \tag{4.2}$$

By analogy, it is possible to defines the apparent efficiency of electricity production in a steam turbogenerator after the unit is repowered:

$$\chi = \frac{N_{el}^{El} + \Delta N_{el}^{ST}}{\dot{E}_{ch}^{coal}} = \frac{N_{el}^{ST}}{\dot{E}_{ch}^{coal}}. \tag{4.3}$$

In terms of annual capacity the equations take the following form:

$$\eta_{Eel,A}^p = \frac{E_{el,A}^{ST} + E_{el,A}^{GT}}{E_{ch,A}^{coal} + E_{ch,A}^{gas}} \tag{4.4}$$

$$\eta_{\Delta,A} = \frac{E_{el,A}^{GT} + \Delta E_{el,A}^{ST}}{E_{ch,A}^{gas}} \tag{4.5}$$

$$\chi_A = \frac{E_{el,A}^{ST}}{E_{ch,A}^{coal}} \tag{4.6}$$

where:

$E_{ch,A}^{coal}$, $E_{ch,A}^{gas}$	annual use of the chemical energy of coal and gas after repowering the unit by a gas turbine and heat recovery steam generator,
$E_{el,A}^{ST}$	annual electrical output of the steam turbogenerator after repowering the unit by a gas turbogenerator and heat recovery steam generator,
$\Delta E_{el,A}^{ST}$	increase of the annual electrical output of the steam turbogenerator in comparison to the output prior to repowering of the unit,
$E_{el,A}^{GT}$	annual production of electricity in the gas turbogenerator.

The values of the variables $E_{ch,A}^{coal}$, $E_{ch,A}^{gas}$, $E_{el,A}^{ST}$, $\Delta E_{el,A}^{ST}$, $E_{el,A}^{GT}$ will be relative of the
capacity of the gas turbine N_{el}^{GT} applied in the power unit and the structure of heat
recovery steam generator.

The thermodynamic criterion for the selection of the capacity gas turbine and
structure of the waste heat boiler for the repowered unit is the maximization of its
efficiency:

$$\eta_{Eel,A}^p = \frac{E_{el,A}^{ST} + E_{el,A}^{GT}}{E_{ch,A}^{coal} + E_{ch,A}^{gas}} \rightarrow max. \tag{4.7}$$

References

1. Bartnik R (2009) Combined cycle power plants. Thermal and economic effectiveness. WNT, Warszawa
2. Bartnik R, Buryn Z (2011) Conversion of coal-fired power plants to cogeneration and combined-cycle. Thermal and economic effectiveness. Springer, London

Chapter 5
Selection of an Optimum Gas Turbogenerator for the Repowered Power Unit

Abstract This chapter presents results of technical calculations involving the selection heating structures of heat recovery steam generators and of power of gas turbine for modernization of coal-fired power unit to dual-fuel steam-gas system. Calculations were conducted for the condensing mode and for operation of the unit in cogeneration. Differences between the optimal thermodynamic capacity of the gas turbine for the condensing mode and cogeneration are presented with regard to a 370 MW power unit and relative to the number of degrees of pressure in the heat recovery steam generator.

Keywords Selection · Gas turbine · Condensing operation of a unit · Cogeneration unit operating

The structure of the gas turbines, their operating parameters and efficiency of the existing gas turbines applied in up-to date power systems cannot be changed. The current state of technology does not enable the design and production of gas turbines which result from the optimization of their specific parameters. Hence, in gas-steam systems in use it is necessary to use the available structures. The only issue is associated with the selection of a adequate turbine to fit the capacity of the system.

For the case of parallel repowering of a unit by a gas turbine—(Chap. 1, Fig. 1.1b, 5.1)—the energy efficiency of such repowering is considerable dependent on the capacity N_{el}^{GT} and structure of a heat recovery steam generator—as it stems from Eqs. (4.1, 4.7). It is, therefore, necessary to optimize the capacity of a turbine and the structure of a waste boiler while selecting them for coupling with a 370 MW power unit.

The capacity of a turbine and the stream of the enthalpy of the exhaust gas fed into heat recovery steam generator are expressed by the equation [1, 2]

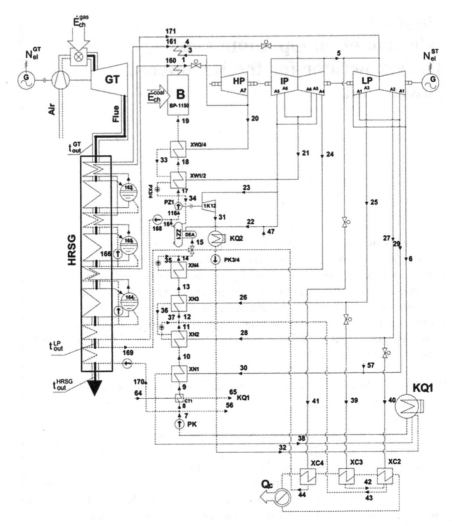

Fig. 5.1 Thermal diagram of 370 MW electric power unit repowered by gas turbine with triple pressure heat recovery steam generator and operating under combined heat and power with steam supply to XC2, XC3, XC4 heaters (*GT* – gas turbine; *HRSG* heat recovery steam generator ; XC2, XC3, XC4—heaters; in the alternative with repowering the system to a dual-fuel combined-cycle without cogeneration XC2, XC3, XC4 heaters are excluded)

$$\dot{H}_{fg} = \frac{N_{el}^{GT}}{\eta_{GT}} - N_{el}^{GT} = \frac{\sum_i \dot{Q}_i}{\eta_{HRSG}}, \tag{5.1}$$

where:

$\dot{Q}_i$—thermal power given off to steam, feedwater, condensate by exhaust gas from the gas turbine in an *i*-th heat exchanger installed in a heat recovery steam generator, η_{GT}—efficiency of the gas turbogenerator, η_{HRSG}—efficiency of the heat recovery steam generator.

From the Eq. (5.1) one can see that the parallel system is distinguished by considerable freedom in terms of selecting the capacity of the gas turbine and use of the enthalpy of the exhaust gas in the heat recovery steam generator. The capacity of the turbine N_{el}^{GT} can be arbitrarily large and is relative to the value of capacity $\dot{Q}_i$. The limitations imposed on it are associated with the economic issues such as financial capabilities of an investor. For thermodynamic reasons the higher the degree in which the gas turbine can relieve the production of the coal-fired boiler, which forms the major sources of exergy losses in the system, the higher the energy efficiency of the production of electricity. The increase in the power of the gas turbine can be limited by the admissible overloading of the blading system of the steam turbine or by the overloading of the electric generator coupled with it. The maximum admissible overloading of the GTHW-370 generator is equal to 406 MW.

The highest overloading of the steam turbine occurs in its low-pressure section LP as a result of feeding higher amounts of steam than for the case of rated loading. During overcharging the bending stress increases in the blades as a consequence of higher aerodynamic forces in the casing and disk and the axial pressure on the bearing. The elements which are most susceptible to failure due to overloading are the blades in the rear stages of the low-pressure section of the LP turbine. It is so as there is high tensile stress from the centrifugal force acting on the blades. Another important consideration in the overloading of the LP section is associated with the possibility of undesirable stagnation of steam flow through its final stages, which could lead to the decrease of its internal efficiency. The stream of exhaust steam from the LP section through the throat into the condenser should not be continuously larger by more than 10 % in comparison to the rated loading. For the 18K370 turbine the rated (reference) flow is equal to 195.1 kg/s, while admissible one—218.2 kg/s. For higher flow rate it would be necessary to modernize the throat of the turbine and the condenser. The maximum steam flow from the LP section to the condenser for the temporal limitation of such operating conditions to 2–3 h per day, can reach around 240 kg/s. This is additionally possible for the exclusion of the high-pressure regeneration. Annually the time for such operation is limited to around 600 h/a.

The maximum admissible pressure in the condenser is equal to 17–18 kPa, despite the fact that the calculated (reference) pressure in it can reach as much as 25 kPa. For this value the protection equipment brings about shut-down of the turbine (then we have to do with the idle work of the boiler). However, problems start to occur from exceeding the pressure of around 18 kPa, which is often possible during the summer season. This leads to an increase of pressure in the regulation wheel of the turbine, whose admissible pressure is 15.2 MPa. In principle, the maximum output is then limited to e.g. 350 MW, which is reported to national Electrical System Operator, who decides on the adequate changes in the loading of a power unit. If the resulting power loss can be supplemented via the use of remaining units in a power plant, no financial losses are recorded in the power plant. However, if it is not possible, they have to purchase the deficit of power on the balancing energy market.

The problem associated with the overloading of the steam turbogenerator is only encountered in a parallel system for a relatively large capacity of a gas turbine (in an in-series system this problem does not occur as the heat recovery steam generator is excluded). In this case the generation of resuperheated steam and low-pressure steam in a heat recovery steam generator for a concurrent reduction of extraction of steam into low-pressure regeneration heaters as a result of its partial substitution in a heat recovery steam generator can lead to a considerable increase of the power output from low-pressure section of the turbine and flow of steam into the condenser. In this case, however, it would be necessary to modernize both this part of the turbine as well as the condenser. Such modernization would, however, be cheaper than the application of a Hot Windbox system, for which the problem of steam turbogenerator overloading does not occur. It is so since in terms of necessary investment will be considerably higher than the expenditure needed for a heat recovery steam generator and modernization of the low-pressure section of the steam turbine in a parallel system.

5.1 Condensing Operation of a Unit

This chapter presents thermodynamic analysis of the conversion of 370 MW unit to dual-fuel gas-steam system with parallel coupling—Fig. 1.1b, 5.1. The analysis involves the conditions of its operation after its repowering by a gas turbogenerator and heat recovery steam generator. The computer calculations involving multiple alternatives apply the mathematical model of a power unit presented in [2]. Calculations were conducted for condensing operation of the unit (XC2, XC3, XC4 heaters—Fig. 5.1—are excluded).

5.1.1 Results of Thermodynamic Calculations

The thermodynamic analysis of the repowered unit is undertaken for the entire range of the capacities of the manufactured gas turbines $N_{el,n}^{GT} \in (0; 350$ MW$)$ and for single-, dual- and triple-pressure heat recovery steam generators—Fig. 5.2. The results of thermodynamic calculations for average annual value of ambient temperature equal to +8, 1 °C are presented in Figs. 5.3, 5.4, 5.5, 5.6, 5.7, 5.8, 5.9, 5.10.

The repowering of a 370 MW unit by a gas turbogenerator and heat recovery steam generator results in a reduced use of coal in a BP-1150 boiler—Fig. 5.3. As a result, to the same degree, exergy losses decrease in it and efficiency of the

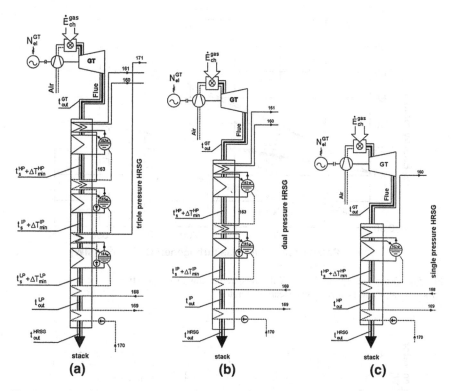

Fig. 5.2 Schematic diagram of gas turbine and heat recovery steam generators; **a** triple pressure *HRSG*, **b** dual pressure *HRSG*, **c** single pressure *HRSG*

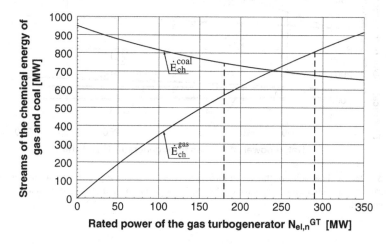

Fig. 5.3 Streams of the chemical energy of gas and coal combustion in an existing BP-1150 coal-fired boiler in the function of the capacity of the gas turbogenerator and structure of heat recovery steam generator (stream of the chemical energy of the gas is not relative to the structure of heat recovery steam generator)

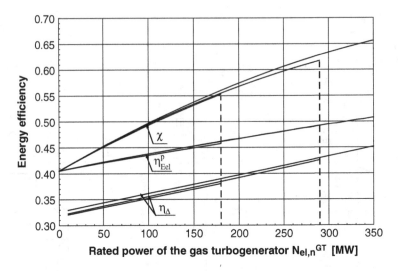

Fig. 5.4 Gross energy efficiency of the repowered unit in the function of the capacity of the turbogenerator and heat recovery steam generator

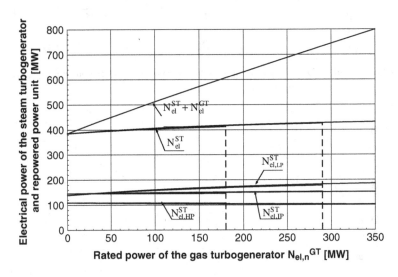

Fig. 5.5 Electrical capacity of the steam turbogenerator and total output of a power unit after its repowering in the function of capacity of the gas turbogenerator and structure of heat recovery steam generator

production of electricity η^p_{Eel} (Chap. 4, Eq. 4.1) increases in a repowered unit—Fig. 5.4. Figure 5.4 also presents the incremental efficiency η_Δ and apparent efficiency χ of the power unit (Chap. 4, Eqs. 4.2, 4.3).

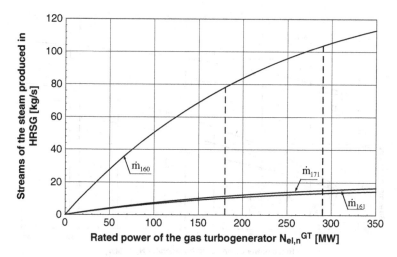

Fig. 5.6 Streams of fresh, intermediate- and low-pressure steam produced in the heat recovery steam generator in the function of the capacity of the gas turbogenerator and structure of the heat recovery steam generator

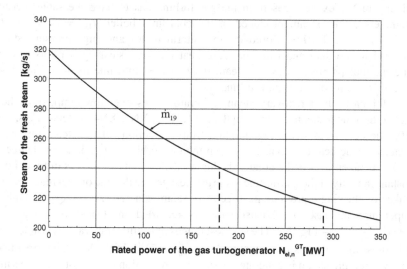

Fig. 5.7 Stream of fresh steam produced in the existing coal-fired boiler BP-1150 in the function of the capacity of the gas turbogenerator and structure of the heat recovery steam generator

The reduced use of coal results from the fact that the heat recovery steam generator takes over the production of fresh steam from the BP-1150 boiler. Concurrently, the overall stream of steam which is fed into the steam turbine remains constant $\dot{m}_1 = \dot{m}_{19} + \dot{m}_{160} = $ const—Figs. 5.6, 5.7, 5.8. The produced stream $\dot{m}_{160}$ in the heat recovery steam generator by means of low-temperature

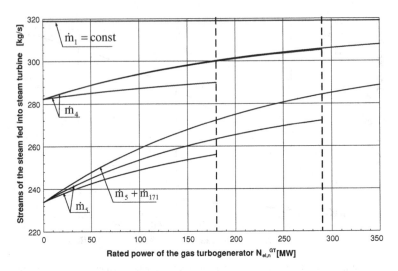

Fig. 5.8 Streams of steam fed into steam turbine in the function of the capacity of the gas turbogenerator and structure of the heat recovery steam generator

enthalpy of the exhaust gas from the gas turbine has to have the same thermal parameters as the steam produced in the coal-fired boiler: $t_{160} = t_1 = 540\ °C$, $p_{160} = p_1 = 18, 3\ MPa$. Moreover, the intermediate- and high-pressure steam produced in the dual and triple-pressure heat recovery steam generators has the same thermal parameters as the steam used to feed into intermediate- and low-pressure sections of the steam turbine.

Additionally, heat recovery steam generator takes over low-temperature heat regeneration in the steam cycle from the XN1, XN2, XN3, XN4 heaters—Figs. 5.9, 5.10. The heat from steam regeneration from extractions of the steam turbine is replaced by the heat of exhaust gas from the gas turbine—Fig. 5.1. The degree in which it is taken over is the greater the higher the capacity of the gas turbine. For a constant capacity of the gas turbine to the greatest degree the role of heat regeneration is taken by the single-pressure heat recovery steam generator. It is so as the highest temperature increase of exhaust gases is recorded in it and it is equal to $\Delta t = t_{out}^{HP} - t_{out}^{HRSG}$—Fig. 5.2. The smallest ranges of useful temperatures is noted in a triple-pressure boiler, as it is equal to $\Delta t = t_{out}^{LP} - t_{out}^{HRSG}$. In the latter, intermediate- and low-pressure steam is generated beside the production of fresh steam (the same volume as in single- and dual-pressure boilers).

The fact that the heat recovery steam generator takes over heat regeneration additionally increases the electrical efficiency of the system, as the exergy losses are limited in it. The reduction of the exergy losses is associated with exhaust streams of steam fed into regenerative preheaters XN1–XN4. These losses are the smaller, the greater the degree in which regeneration takes over the role of heat recovery steam generator. The steam which is unnecessary for the purposes of regeneration is then fed into the condenser KQ1 and is being expanded

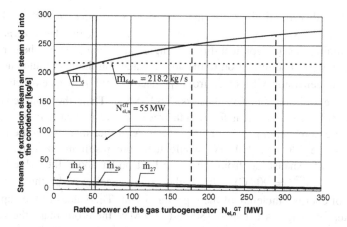

Fig. 5.9 Streams of extraction steam fed into low-pressure regenerative preheaters XN1, XN2, XN3 and stream of steam fed into the condenser in the function of the capacity of the gas turbogenerator and structure of the heat recovery steam generator

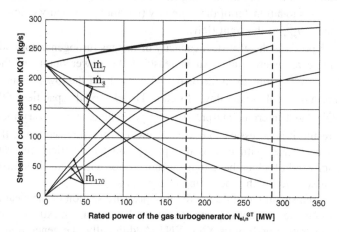

Fig. 5.10 Streams of condensate from KQ1 condenser and fed into low-pressure regenerative preheaters XN1, XN2, XN3, XN4 and regenerative preheater in the heat recovery steam generator in the function of the capacity of the gas turbogenerator and structure of the heat recovery steam generator

adiabatically in the steam turbine. As a result, the capacity of the turbogenerator increases (even more than the exergy streams from this steam as the pressure in the condenser is lower than the ambient pressure). Concurrently, the exergy losses as a result of heat flow in the regenerative preheater in the heat recovery steam generator is smaller than the avoided exergy losses of the extraction steam due to the low mean thermodynamic temperature of the gas; therefore, the electrical efficiency of the repowered unit increases.

The dashed vertical lines for the electrical capacities of $N_{el,n}^{GT} = 180$ and 290 MW Figs. 5.3, 5.4, 5.5, 5.6, 5.7, 5.8, 5.9, 5.10 mark the thermodynamically justified limitations for applicability of single- and dual-pressure heat recovery steam generators. The limitation to the above values on the maximum capacity of the gas turbogenerator $N_{el,n}^{GT}$ stems from the decay of the flow of extraction steam into low-pressure regenerative heaters XN1–XN4. Then the total condensate stream from KQ1 condenser is fed into the regenerative preheater in the of single- and dual-pressure steam generator—Fig. 5.10. A further increase in the electrical capacity above 180 and 290 MW would therefore result in an increase of the temperature of exhaust gas from the waste heat boiler above the value taken for calculations ($t_{out}^{HRSG} = 90$ °C). The higher the capacity $N_{el,n}^{GT}$, the greater the decrease of energy efficiency of the repowered unit. The values of the capacity of the gas turbogenerators equal to 180 and 290 MW therefore mark the optimum values in terms of thermodynamic efficiency for the subsequent single- and dual-pressure heat recovery steam generators applied in the system. For a triple-pressure steam generator the value above 350 MW is optimum capacity of the gas turbogenerator. An assumption of a higher (lower) value of the temperature t_{out}^{HRSG} from 90 °C leads to a transfer of the boundary (thermodynamically efficient) value of the capacity $N_{el,n}^{GT}$ to the right (left).

Figure 5.5 presents the electrical capacity of the gas turbogenerator and overall capacity of the repowered unit. An increase of the power output of the steam turbogenerator N_{el}^{ST} results from the increased capacity of the low-pressure section $N_{el,LP}^{ST}$ of it as a result of greater flow of steam. The electrical capacities of the high-pressure $N_{el,HP}^{ST}$ and intermediate-pressure $N_{el,IP}^{ST}$ sections tend to vary inconsiderably. As a result, it is necessary to design a new low-pressure section of the steam turbine and condenser with increased capacity and an electrical generator with bigger capacity, all of which have to be accounted for during considerations of the investment associated with repowering of the unit.

The increased steam feed through the low-pressure section of the steam turbine and condenser comes as a consequence of smaller heating steam extraction into low-pressure regenerative preheaters XN1–XN4. Additionally, the generation of intermediate- and low-pressure steam in the heat recovery steam generator—Figs. 5.6, 5.8—contributes towards an increase of the capacity of the steam turbogenerator.

For the case of repowering a 370 MW unit by a gas turbogenerator with the capacity of $N_{el,n}^{GT} = 350$ MW and a triple-pressure heat recovery steam generator, the overall capacity of the repowered unit reaches as much as 800 MW.

If the capacity of a gas turbogenerator does not exceed 55 MW, the stream of steam fed into condenser $\dot{m}_6$—Fig. 5.9—does not exceed the admissible value of 218.2 kg/s and it is not necessary to redesign the low-pressure section LP of the steam turbine and the condenser KQ1 and, as a consequence, new electrical generator is not needed either. From the multitude of alternatives considered in the calculations, the maximum admissible stream of steam fed into the condenser forms the most severe limitation deciding on whether it is necessary to replace the low-pressure section of the steam turbine, condenser and the electric generator. No limitations in this respect, as it results from the calculation, are associated with the pressures in the turbine extractions, since the admissible values resulting from the increased flow of steam are not exceeded in them.

5.2 Cogeneration Unit Operating

The most effective way of reducing the consumption of chemical energy of the fuels and emission of harmful products during fuel combustion into the environment is the cogeneration, i.e. concurrent production of heat and electricity. The adaptation of a power unit to cogeneration mode will result in the improvement of the total energy efficiency of its operation. Moreover, its simultaneous super structuring to cogeneration mode will increase this effectiveness to a bigger extent [1, 2]. Modernization of a power unit by repowering it with a gas turbogenerator and the heat recovery steam generator also leads to the improvement of its economic effectiveness. Mainly, this is dependent on the relations of prices between energy carriers, i.e. on price relations of between heat and electricity and fuel prices, coal and gas as well as on the overall capacity of the system, i.e. amount of electricity and heat production in the unit and thereby, on the capacity of a gas turbogenerator and the structure of a waste-heat boiler used to super structure the power plant.

Section 5.1 presents a thermodynamic analysis of the condensing mode of a 370 MW power unit repowered by a gas turbogenerator and a heat recovery steam generator. In this its operation in cogeneration mode is analyzed—Fig. 5.1. Differences between the optimal thermodynamic capacity of the gas turbine for the condensing mode and cogeneration are presented with regard to a 370 MW power unit and relative to the number of degrees of pressure in the waste-heat boiler.

5.2.1 Results of Thermodynamic Calculations

The analysis has been conducted, as in Sect. 5.1, for the entire range of the capacities of manufactured gas turbines $N_{el,n}^{GT} \in (0; 350 \text{ MW})$ and for single-, dual- and triple-pressure heat recovery steam generators—Fig. 5.2.

The operation of a power unit under cogeneration can be characterized by a big fluctuations of steam bleed from A2, A3 extractions and IP-LP crossover pipe in the steam turbine used to feed heat exchangers XC2, XC3 and XC4 —Fig. 5.11 [2]. These fluctuations originate from the variable demand for thermal power and its quality regulation—Figs. 5.12, 5.13 (on Fig. 5.12 heating power $\dot{Q}_{dhw}$ has not been indicated for the needs of domestic hot water preparation; this power is delivered to the customers both in a peak season and off-peak season).

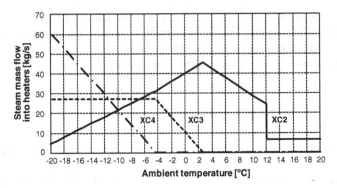

Fig. 5.11 Streams of extraction steam into XC2, XC3, XC4 heaters in the function of ambient temperature

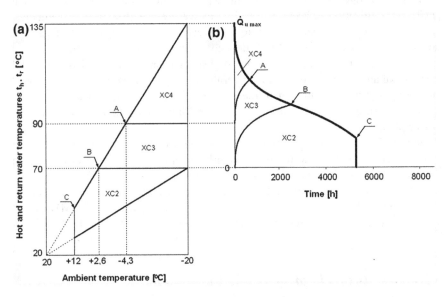

Fig. 5.12 Qualitative regulation of thermal power output $\dot{Q}_u$ from the power station for the purposes of heating, air conditioning and ventilation of residential areas for the alternative with three heaters XC2, XC3 and XC4: **a** linear regulation chart; **b** annual scheduled chart of demand for thermal power (t_h, t_r—temperatures of network hot water and return water)

The stream of the heating steam is subjected to changes depending on the ambient temperature. For example, steam extracted from a IP-LP crossover pipe to feed XC4 heater assumes the biggest value at the peak season for district heating and zero value at point A and to the right of it —Fig. 5.12b. At this point and on its left the steam from A3 extraction to feed XC3 heater keeps at a permanently high value, while at point B it is equal to zero, and the steam from A2 extraction to feeding XC2 heater at this point assumes its maximum value. The smallest stream

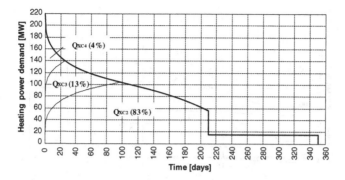

Fig. 5.13 Annual scheduled demand for heating power

of heating steam stream comes from A2 extraction it is used exclusively for the needs of producing domestic hot water. In the XC2 heater network water heated to the temperature of $t_h = 70\ °C$, while in XC3 heater to the temperature of $t_h = 90\ °C$ and in XC4 heater to the temperature of $t_h = 135\ °C$—Fig. 5.12a.

The peak thermal capacity of $\dot{Q}_{c\,max} = \dot{Q}_{u\,max} + \dot{Q}_{dhw} = 220$ MW was adopted in the calculations and the power of $\dot{Q}_{dhw} = 15$ MW for the purposes of producing domestic hot water—Fig. 5.13.

The selected outcomes of thermodynamic calculations involving several alternatives are presented in Figs. 5.14, 5.15, 5.16. The results are presented for three ambient temperatures: $-20\ °C$, $+8, 1\ °C$ and $+20\ °C$. It is purposeful as at these temperatures the power unit operates at different thermal powers so the values of particular thermodynamic parameters of its operations are different. Hence, there are different conditions resulting from them, which decide on the boundary capacity of the gas turbine power used to repower the existing coal-fired power unit. At the temperature of $-20\ °C$ the power unit operates with the maximum heating power of 220 MW; thus, with the maximum stream of bleed steam into XC2, XC3 and XC4 heaters. The thermal power for an average annual temperature of $+8, 1\ °C$ is considerably smaller than the maximum power, and for the temperature of $+20\ °C$ the power unit works with the lowest annual heating power equal to 15 MW only for the purposes of domestic hot water—Figs. 5.11, 5.12, 5.13.

Broken vertical lines in Figs. 5.14, 5.15, 5.16 used to limit the particular analytic curves come, just as in Sect. 5.1, from the decay of heating steam bleed into the heaters in the section of low-pressure regeneration. Then low pressure regeneration is "taken over" by the exhaust gas-water heater situated in the rear section of the heat recovery steam generator for the range of low temperature exhaust gases. The further increase in the capacity of the gas turbine would therefore result in the decrease of energy efficiency of the modernized power unit. Then the temperature of the flue gas from the heat recovery steam generator would increase above the temperature adopted for its calculations $t_{out}^{HRSG} = 90\ °C$. The coordinate abscissa of the broken lines therefore represent the optimal

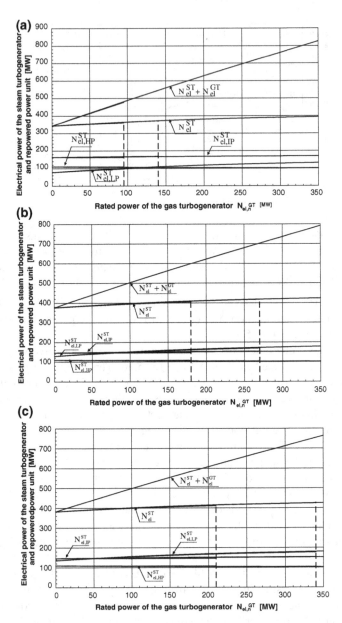

Fig. 5.14 a Electrical capacity of the steam turbogenerator and total output of a power unit after its repowering in the function of the capacity of the gas turbogenerator and structure of the heat recovery steam generator for ambient temperature $t_{amb} = -20\ °C$. **b** Electrical capacity of the steam turbogenerator and total output of a power unit after its repowering in the function of the capacity of the gas turbogenerator and structure of the heat recovery steam generator for ambient temperature $t_{amb} = +8{,}1\ °C$. **c** Electrical capacity of the steam turbogenerator and total output of a power unit after its repowering in the function of the capacity of the gas turbogenerator and structure of the heat recovery steam generator for ambient temperature $t_{amb} = +20\ °C$

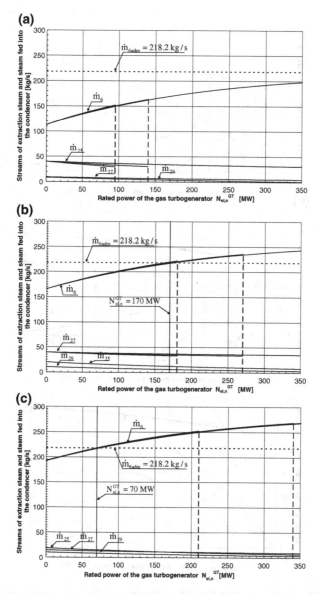

Fig. 5.15 a Streams of extraction steam fed into low-pressure regenerative preheaters XN1, XN2, XN3 and stream of steam fed into the condenser in the function of the capacity of the gas turbogenerator and structure of the heat recovery steam generator for ambient temperature $t_{amb} = -20\ °C$. **b** Streams of extraction steam fed into low-pressure regenerative preheaters XN1, XN2, XN3 and stream of steam fed into the condenser in the function of the capacity of the gas turbogenerator and structure of the heat recovery steam generator for ambient temperature $t_{amb} = +8,1\ °C$. **c** Streams of extraction steam fed into low-pressure regenerative preheaters XN1, XN2, XN3 and stream of steam fed into the condenser in the function of the capacity of the gas turbogenerator and structure of the heat recovery steam generator for ambient temperature $t_{amb} = +20\ °C$

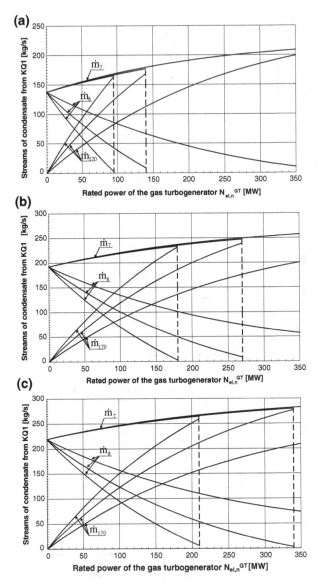

Fig. 5.16 a Streams of condensate from KQ1 condenser and fed into low-pressure regenerative preheaters XN1, XN2, XN3, XN4 and regenerative preheater in the heat recovery steam generator in the function of the capacity of the gas turbogenerator and structure of the heat recovery steam generator for ambient temperature $t_{amb} = -20$ °C. **b** Streams of condensate from KQ1 condenser and fed into low-pressure regenerative preheaters XN1, XN2, XN3, XN4 and regenerative preheater in the heat recovery steam generator in the function of the capacity of the gas turbogenerator and structure of the heat recovery steam generator for ambient temperature $t_{amb} = +8,1$ °C. **c** Streams of condensate from KQ1 condenser and fed into low-pressure regenerative preheaters XN1, XN2, XN3, XN4 and regenerative preheater in the heat recovery steam generator in the function of the capacity of the gas turbogenerator and structure of the heat recovery steam generator for ambient temperature $t_{amb} = +20$ °C

thermodynamic capacities of gas turbines for single- and dual-pressure waste heat boilers. Moreover, they vary in accordance with the ambient temperature, in distinction to the condensing operation of the power unit (Sect. 5.1). For the ambient temperature -20 °C in case of single-pressure waste boiler the maximum thermodynamically justified (optimal) capacity of the gas turbogenerator is equal to $N_{el,n}^{GT} = 95$ MW, and in the case of dual-pressure boiler it is $N_{el,n}^{GT} = 140$ MW— Fig. 5.16a. For these capacities the entire stream of the condensate from condenser KQ1 is fed into the regeneration heater, for the single- and dual-pressure boilers, respectively. Thus, the further increase of the capacity of the gas turbines above 95 and 140 MW, respectively for case of single- and dual-pressure boilers, would further increase the temperature of exhaust gases from the boilers above the adopted temperature $t_{out}^{HRSG} = 90$ °C, so the energy efficiency of the repowered unit would decrease. For the temperature of +8, 1 °C these capacities are equal to 180 and 270 MW, respectively, as in Fig. 5.16b, while for the temperature of +20 °C, the respective values are 210 and 340 MW, Fig. 5.16c. For the triple-pressure heat recovery steam generator the optimal capacity of the gas turbogenerator is over 350 MW, regardless of the ambient temperature.

The smallest boundary value of the capacity of the gas turbogenerator is imposed by the operation under cogeneration in the off-peak (summer) season with the thermal capacity of 15 MW. This boundary value is equal to $N_{el,n}^{GT} = 70$ MW—Fig. 5.15c. Above this capacity, it is already necessary to install a new low-pressure section LP of the steam turbine, the condenser KQ1 and the electric generator, Fig. 5.1, with higher capacities. If the power unit operated over the entire year with the thermal power of 220 MW, it would be unnecessary to interfere with the structure of the existing steam turbine—Figs. 5.14a, 5.15a. During operation with average annual heating power for the temperature of +8, 1 °C, the boundary capacity of the gas turbogenerator is equal to 170 MW— Fig. 5.15b. Therefore, in practice the operating conditions in off-peak season with heating power for the needs of domestic hot water decides on the necessary extent of modernization of the low-pressure LP section of the power unit and the installation of a new electric generator. However, the capacities of the high-pressure HP and intermediate-pressure IP basically do not change, just as in the case of condensational work of power unit (Sect. 5.1). The capacities of the specific facilities are presented in Fig. 5.14a, b, c.

As it has been stated, the increase of the capacity of the steam turbogenerator after repowering the unit by the gas turbogenerator with the capacity of above $N_{el,n}^{GT} = 70$ MW is a result of the increase of the capacity of its low-pressure section LP caused by the greater flow of steam despite heating steam bleed into XC2, XC3, XC4 heaters. Therefore, it is necessary to design a new low-pressure section of the steam turbine, a condenser KQ1 and a new electric generator with a greater capacity.

The increased flow of steam through the low-pressure section LP of the steam turbine and the condenser and following increase of its capacity, comes a consequence of the decreased steam bleed fed into low-pressure regeneration heaters

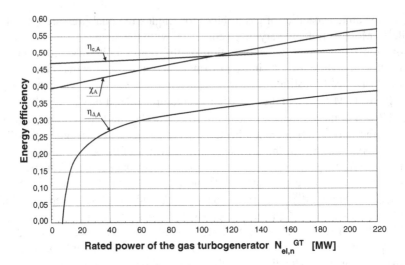

Fig. 5.17 Mean annual energy efficiency of the repowered 370 MW unit in the function of the capacity of the gas turbogenerator coupled with dual-pressure heat recovery steam generator

XN1, XN2, XN3, XN4 (Fig. 5.15a, b, c) due to its partial substitution by regeneration in heat recovery steam generator and due to the smaller flow of condensate from the condenser KQ1 to them (Fig. 5.16a, b, c). This smaller flow of condensate from the condenser into XN1–XN4 heaters results from the change in the regeneration as well as the use of steam for the needs of district heating, as discussed before. Besides, the production of intermediate- and low-pressure steam, in case of application of dual- and triple-pressure waste heat boilers in system, contributes to the increase of capacity of the steam turbogenerator (compare Sect. 5.1.1).

If the power of gas turbine does not exceed 70 MW, steam fed into the condenser $\dot{m}_6$ (Fig. 5.15a, b, c) does not exceed the admissible value equal to 218.2 kg/s (Chap. 5) and it is not necessary to install either new low-pressure section LP of the steam turbine and the condenser KQ1. Besides, a new electric generator is not necessary. As we can conclude from the conducted calculations, the maximum tolerated steam bleed $\dot{m}_6$ to feed the condenser is forms the strictest limitation deciding whether on the potential need to exchange the low-pressure section of the steam turbine, condenser and electric generator.

Apart from the change of power of LP part of steam turbine and the increase of total electrical capacity of the unit, there is a considerable increase of its energy efficiency—Fig. 5.17. The detailed calculations of only average annual values of efficiency (and not for the particular ambient temperatures of −20 °C, +8, 1 °C and +20 °C) have been undertaken exclusively for the dual-pressure waste heat boiler. Only such a waste heat boiler is economically justified (Eq. 6.3). Furthermore, the calculations for the range of gas turbine capacity of 220 MW have been limited because of the decreasing heating steam bleed into the low pressure regeneration XN1, XN2, XN3, XN4 heaters. The further increase of the capacity would be thermodynamically unjustified.

The negative value of the incremental efficiency $\eta_{\Delta, A}$ (being the equivalent of the efficiency of generating electricity in the single-fuel gas-steam system) for the capacity of the gas turbogenerator within the range below about 8 MW is not physically contradictory, in accordance with the definition in [1, 2]. The increase in the capacity of the steam turbine as a result of cogeneration is in this case negative and the absolute value is greater than the power of the gas turbogenerator.

While the power unit is being repowered by a gas turbogenerator with the capacity of 350 MW and by a triple-pressure heat recovery steam generator, the power unit's electrical capacity increases even two times and is equal to about 800 MW—Fig. 5.14a, b, c—despite the cogeneration (heat production is then relatively small, even during the peak period, in relation to the production of electricity). Thus, the total energy efficiency of the power unit is equal to (compare Chap. 4, Eq. 4.4)

$$\eta_{c,A} = \frac{E_{el,A}^{ST} + E_{el,A}^{GT} + Q_{c,A}}{E_{ch,A}^{coal} + E_{ch,A}^{gas}} \simeq 60\ \%, \tag{5.2}$$

the incremental efficiency amounts to about $\eta_{\Delta, A} \approx 39\ \%$, and the apparent efficiency of the steam turbogenerator is as much as $\chi_A \approx 66\ \%$ (Chap. 4, Eqs. 4.5, 4.6) [1, 2]. The apparent efficiency is to some extent equivalent to efficiency of electricity production in the power unit before its repowering and is equal to 41 % gross.

5.3 Summary and Conclusions

For the case of the 370 MW unit repowered by gas turbogenerator and adapted to cogeneration, the boundary capacity of the gas turbogenerator is equal to 70 MW. Above this capacity the stream of steam fed into the condenser $\dot{m}_6$ exceeds the admissible value of 218.2 kg/s, regardless of the number of pressure stages in the waste heat boiler, Fig. 5.15c, so it is necessary to install low-pressure steam turbine LP with a greater capacity along with a condenser KQ1 and a new electric generator is required. In case of the condensing operation of the power unit, which is associated with no heating steam bleed from the steam turbine into XC2, XC3, XC4 heaters, the boundary capacity of the gas turbogenerator is 55 MW (Sect. 5.1).

Another difference between the operation of a power unit under condensing cycle and cogeneration is the difference in the values of temperatures of the exhaust gases from single-, dual- and triple-pressure waste heat boilers relative to the ambient temperature. In case of a cogeneration cycle, due to heating steam bleed into XC2, XC3, XC4 heaters, this variability is relatively big, which results from the changes of the fluxes $\dot{m}_6$, $\dot{m}_7$ and consequently $\dot{m}_{170}$—Fig. 5.16a, b, c.

References

1. Bartnik R (2009) Combined cycle power plants. Thermal and economic effectiveness. WNT, Warszawa
2. Bartnik R, Buryn Z (2011) Conversion of coal-fired power plants to cogeneration and combined-cycle. Thermal and economic effectiveness. Springer, London

Chapter 6
Selection of the Structure of the Heat Recovery Steam Generator for the Repowered Power Unit

Abstract This chapter presents the methodology of analyzing thermodynamic and economic effectiveness for the selection of a structure of the heat recovery steam generator for the repowered power unit. The calculations were conducted for the unit with the rated capacity of 370 MW.

Keywords Selection · Heat recovery steam generator · Number of degrees of pressure · Thermodynamic effectiveness · Economic effectiveness

Due to the decreased use of coal in the BP-1150 boiler—Chap. 5, Fig. 5.3—to the same extent the loss of the unusable higher range of the temperatures, that is range between the temperature of the combustion of coal in the boiler and temperature of the steam produced in it, is reduced as well. Consequently, the efficiency of the production of electricity in the repowered unit increases as well. The greater the decrease in the use of coal, the higher the required capacity of the gas turbo-generator, as a consequence of this, there are smaller exergy losses and the efficiency of the production of electricity increases as well—Chap. 5, Fig. 5.4.

The reduction of the losses in generation of electricity in a repowered unit can additionally result from the decrease of the exergy losses in the heat recovery steam generator, which is a device coupling the Joule cycle of the gas turbine with the Clausius-Rankine cycle of the steam turbine.

The exergy losses in the heat recovery steam generator stem from the irreversible flow of heat between the gas and water and the steam, and in respect to a unit of time they are expressed by the equation [1]:

$$\delta \dot{B} = T_{amb} \sum \int d\dot{S}_{med} = T_{amb} \int (d\dot{S}_{H_2O} + d\dot{S}_{fg})$$
$$= T_{amb} \int \left(\frac{d\dot{Q}}{T_{H_2O}} - \frac{d\dot{Q}}{T_{fg}} \right) = \int (\theta_{fg} - \theta_{H_2O}) \, d\dot{Q} \tag{6.1}$$

R. Bartnik, *The Modernization Potential of Gas Turbines in the Coal-Fired Power Industry*, SpringerBriefs in Applied Sciences and Technology, DOI: 10.1007/978-1-4471-4860-9_6, © The Author(s) 2013

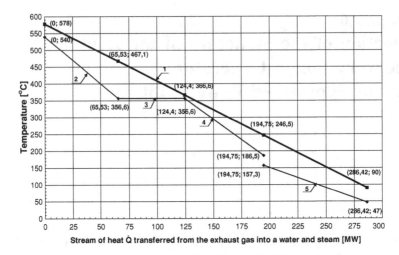

Fig. 6.1 Composition curves in a single-pressure heat recovery steam generator. *1* composition curve for the exhaust gas, *2–5* composition curves for water and steam, *2* primary superheat-ter, *3* high-pressure evaporator, *4* economizer, *5* regenerative preheater

where $\Theta(\Theta = 1 - T_{amb}/T)$ is the exergetic temperature (compare Chap. 2, Eq. 2.1) of exhaust gas and water and steam in the heat recovery steam generator, and T denotes their absolute temperature.

In order to reduce these losses, it is necessary to bring the composition curves of exhaust gases and circulating medium (*pinch method* [1, 2]) in the heat recovery steam generator to similar values by applying several pressure stages, Figs. 6.1, 6.2, 6.3. The losses are minimized as a result of increasing the number of contractions, Figs. 6.4, 6.5, 6.6. As a consequence of increasing the number of pressure stages in a boiler, the difference between the composition curves for exhaust gas and water and steam becomes smaller and, eventually, disappears. However, the investment in the heat recovery steam generator and the remaining components of the steam based section of the gas-steam system increases as well; therefore, the overall capital cost of the power plant increases, too. There is, therefore, a need to search for a technical and economic optimal solution. In practice, in a heat recovery steam generator there are no more than three pressure stages.

As noted above, the thermodynamic criterion for the selection of an optimum structure of a heat recovery steam generator is the one of minimization of the losses of exergy stream during irreversible heat flow between the exhaust gas and circulating media (i.e. water and steam). By integrating Eq. (6.1) (under the assumption that $T = a\dot{Q} + b$), we obtain [1] (compare Chap. 2, Eq. 2.20):

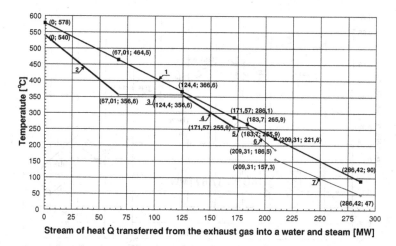

Fig. 6.2 Composition curves in a dual-pressure heat recovery steam generator. *1* composition curve for the exhaust gas, *2–7* composition curve for water and steam, *2* primary and interstage superheater in the 2nd section, *3* high-pressure evaporator, *4* high-pressure economizer and interstage superheater in the 1st section, *5* intermediate-pressure evaporator, *6* intermediate-pressure economizer, *7* regenerative preheater

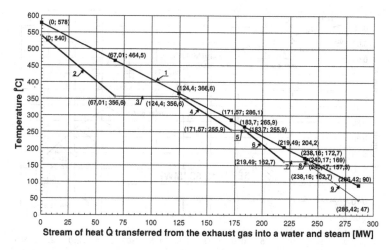

Fig. 6.3 Composition curves in a triple-pressure heat recovery steam generator. *1* composition curve for the exhaust gas, *2–9* composition curve for water and steam, *2* primary and interstage superheater in the 2nd section, *3* high-pressure evaporator, *4* high-pressure economizer and interstage superheater in the 1st section, *5* intermediate-pressure evaporator, *6* intermediate-pressure economizer and low-pressure superheater, *7* low pressure evaporator, *8* low-pressure economizer, *9* regenerative preheater

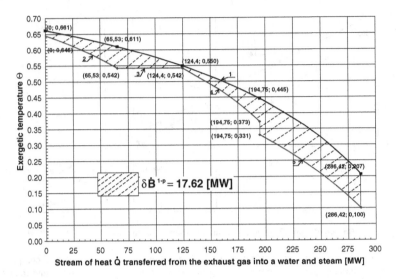

Fig. 6.4 Losses of exergy stream in a single-pressure heat recovery steam generator. *1* composition curve for the exhaust gas, *2–5* composition curves for water and steam, *2* primary superheater, *3* high-pressure evaporator, *4* economizer, *5* regenerative preheater

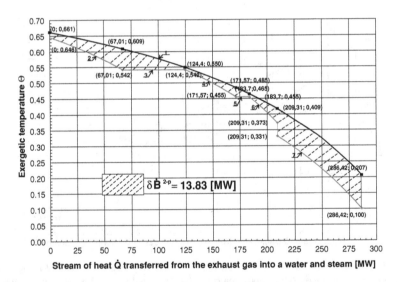

Fig. 6.5 Losses of exergy stream in a dual-pressure heat recovery steam generator. *1* composition curve for the exhaust gas, *2–7* composition curves for water and steam, *2* primary and interstage superheater in the 2nd section, *3* high-pressure evaporator, *4* high-pressure economizer and interstage superheater in the 1st section, *5* intermediate-pressure evaporator, *6* intermediate-pressure economizer, *7* regenerative preheater

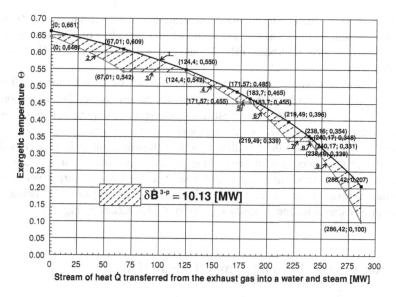

Fig. 6.6 Losses of exergy stream in a triple-pressure heat recovery steam generator. *1* composition curve for the exhaust gas, *2–9* composition curve for water and steam, *2* primary and interstage superheater in the 2nd section, *3* high-pressure evaporator, *4* high-pressure economizer and interstage superheater in the 1st section, *5* intermediate-pressure evaporator, *6* intermediate-pressure economizer and low-pressure superheater, *7* low pressure evaporator, *8* low-pressure economizer, *9* regenerative preheater)

$$
\delta \dot{B} = T_{amb} \sum_i \int_{\dot{Q}_i}^{\dot{Q}_{i+1}} \left(\frac{1}{T_{H_2O}} - \frac{1}{T_{fg}} \right) d\dot{Q}
$$

$$
= T_{amb} \sum_j \ln \sqrt[a_{fg}]{\frac{a_{fg}\dot{Q}_j + b_{fg}}{a_{fg}\dot{Q}_{j+1} + b_{fg}}} \sqrt[a_{H_2O,j}]{\frac{a_{H_2O,j}\dot{Q}_{j+1} + b_{H_2O,j}}{a_{H_2O,j}\dot{Q}_j + b_{H_2O,j}}} \quad (6.2)
$$

$$
+ T_{amb} \sum_k \left(\frac{\dot{Q}_{k+1} - \dot{Q}_k}{T_s^k} - \ln \sqrt[a_{fg}]{\frac{a_{fg}\dot{Q}_{k+1} + b_{fg}}{a_{fg}\dot{Q}_k + b_{fg}}} \right) \rightarrow \min,
$$

where:

i denotes the number of heaters in a heat waste boiler and the difference $\Delta \dot{Q}_i = \dot{Q}_{i+1} - \dot{Q}_i$ denotes the stream of heat exchange in an i-th heater between the exhaust gas with the absolute temperature of T_{fg} and water and steam with the absolute temperature of T_{H_2O}, "i = j + k" (j—number of water superheaters and preheaters; k—number of pressure stages, of the evaporators with the absolute saturation temperature of T_s^k).

The thermodynamic criterion for the selection and arrangement of the heating surfaces should be the one aimed at minimization of the total exergy losses in the boilers: the newly designed heat recovery steam generator and the existing coal-fired boiler BP-1150. The following considerations have to accounted for as well: admissible changes in its loading, and admissible overloading of the existing three-stage, reactive steam turbine 18K370 and the electric generator GTHW-370 coupled with it.

In the heat recovery steam generator it is necessary to consider the installation of such surfaces for which the Eq. (6.2) has the lowest value. The reduction of exergy losses will increase the investment and, as a consequence, the economic effectiveness of the operation of the repowered unit will decrease.

The analysis involves the installation of the following heaters in a waste boiler: surfaces for the generation of high-, intermediate- and low-pressure steam, surface for an inter-stage superheater and a surface for low-pressure regeneration. The thermal parameters of the steam and water are imposed by the existing coal firing power unit with the capacity of 370 MW [3]—Chap. 5, Fig. 5.1.

6.1 Results of Thermodynamic Calculations

The following Figs. 6.1, 6.2, 6.3 present composition curve for the exhaust gas and water and steam in the waste boilers—Chap. 5, Fig. 5.2—for feeding them by exhaust gas from the gas turbogenerator SGT6 with the rated electrical capacity of $N_{el,n}^{GT} = 202$ MW, rated temperature of exhaust gas of $t_{out,n}^{GT} = 578$ °C and rated efficiency of the production of electricity of $\eta_{GT,\ n} = 38.1$ % $(a_{fg} = -\eta_{GT,n} \left(t_{out,n}^{GT} - t_{amb,n} \right) / \left[N_{el,n}^{GT} (1 - \eta_{GT,n}) \right]$, [1], $t_{amb,\ n} = 15$ °C). The temperature of the exhaust gas from the waste heat boiler of $t_{out}^{HRSG} = 90$ °C is taken for these considerations. Figs. 6.4, 6.5, 6.6 present the exergy losses in these boilers (Eq. 6.2).

The losses of exergy stream in a single-pressure heat recovery steam generator is equal to $\delta \dot{B}^{1-p} = 17.62$ MW, in a dual-pressure one—$\delta \dot{B}^{2-p}= 13.83$ MW and in a triple-pressure one—$\delta \dot{B}^{3-p} = 10.13$ MW. The electric power output of the steam turbogenerator will be greater by the value of the difference of losses (under the assumption of a lack of losses during the conversion of mechanical energy into electric energy) in the specific boilers (as indicated in Eqs. Chap. 2, 2.20, 2.21). For instance, by applying a triple-pressure boiler instead of the single-pressure one, the power output from the steam turbogenerator will be higher by $\Delta(\delta \dot{B}^{1-3}) = \delta \dot{B}^{1-p} - \delta \dot{B}^{3-p} = 17.62 - 10.13 = 7.49$ MW. Finally, the selection of a specific heat recovery steam generator in a system should be based on the economic criterion. The necessary condition is associated with the increase of the revenues from the sales of additional amount of electricity which has to be greater than the increase of the annual capital cost and cost of maintenance and overhaul associated with the greater investment ΔJ^{mod} in the repowered unit:

$$\Delta(\delta \dot{B}) e_{el} \tau_A \geq (z \rho + \delta_{serv}) \Delta J^{mod} \qquad (6.3)$$

where:

e_{el} specific sales price of electric energy (per energy unit),

$z\rho + \delta_{serv}$ annual rate of investment service and remaining fixed cost relative to capital expenditure [1, 3, 4],

τ_A time of annual operation of the power plant.

This can ensure that the improvement of the thermodynamic parameters of the modernized power unit as a result of installing greater number of pressure stages in a heat recovery steam generator is cost-effective.

References

1. Bartnik R (2009) Combined cycle power plants. Thermal and economic effectiveness. WNT, Warszawa
2. Szargut J, Ziębik A (2000) Fundamentals of thermal power industry. PWN, Warszawa
3. Bartnik R, Buryn Z (2011) Conversion of coal-fired power plants to cogeneration and combined-cycle. Thermal and economic effectiveness. Springer, London
4. Bartnik R (2008) Technical and economic efficiency account in utility power engineering. Oficyna Wydawnicza Politechniki Opolskiej, Opole

Chapter 7
Comparison of Specific Cost of Producing Electricity in a 370 MW Power Unit Adapted to Dual-Fuel Gas–Steam System and in a New One Operating Under Supercritical Parameters

Abstract This chapter presents the results of economic calculations of the specific cost of electricity production in an existing 370 MW power unit adapted to a dual-fuel gas-steam system and in a new one for supercritical parameters. The calculations involved a power unit with the capacity of 800 ÷ 900 MW.

Keywords Comparison · Specific cost of producing electricity

A technologically and technically rational opportunity for the modernization of existing coal-fired power plants, thus offering their modernization is associated with a conversion into dual-fuel gas-steam systems [1, 2]. The dual-fuel technology offers an additional benefits of modernizing the power engineering sector with the lowest possible expenditure. Such adaptation is four times less expensive per specific unit of the electrical capacity (it is estimated at 1.6 million PLN/MW) in comparison to the construction of a power unit for supercritical parameters (the investment is estimated at 6.5 million PLN/MW). In comparison, nuclear power engineering is 12 times more expensive in investment, which is estimated at 18 million PLN/MW. One can note here that the construction of new power plants for supercritical parameters and nuclear power plants does not enable the modernization of the existing power units.

In the consideration of the above, it is necessary to find an answer to the following. Is it economically justified to modernize the existing coal-fired power units to dual-fuel gas-steam systems, which; however, burn expensive natural gas? Is it more justified to undertake the construction of new expensive power units for supercritical parameters which are fired by the relatively cheap coal? What are the justified relations between fuel prices and what ranges should they take in order to guarantee high economic efficiency of the examined technologies?

R. Bartnik, *The Modernization Potential of Gas Turbines in the Coal-Fired Power Industry*, SpringerBriefs in Applied Sciences and Technology, DOI: 10.1007/978-1-4471-4860-9_7, © The Author(s) 2013

7.1 Specific Cost of Electricity Production in a Power Plant

The specific cost of electricity production in a power plant is expressed by the equation. [1, 3]

$$k_{el} = \frac{K_A}{E_{el,A}} = \frac{K_e + K_{cap}}{E_{el,A}}.$$ (7.1)

where $E_{el,A}$ denotes the annual net production of electricity in a power plant, K_A—total annual operating cost of a power plants made up by the total of exploitation cost K_e and capital cost K_{cap}.

The annual exploitation cost K_e of a power plants involves the cost of fuel purchase and cost of the power plant internal load K_{fuel}, cost of supplementing water K_{sw}, cost of payroll including overheads K_{pay}, maintenance and overhaul cost K_{serv}, cost associated with the purchase of non-energy resources an auxiliary materials K_{am}, charges for the use of environment K_{env} (including emission charges, wastewater disposal, waste storage, etc.), taxes and insurance $K_{tax + ins}$

$$K_e = K_{fuel} + K_{sw} + K_{pay} + K_{serv} + K_{am} + K_{env} + K_{tax+ins}.$$ (7.2)

The cost K_e does not include the cost of the purchase of carbon dioxide emission allowances. This cost can be considerable and even double the specific cost of electricity production in a coal-fired power plant.

The annual capital cost K_{cap} is formed by the total of depreciation and financial cost, i.e. the cost which should offer the return for the investment J with interest from the capital [1, 3]

$$K_{cap} = z\rho J$$ (7.3)

where:

z coefficient of freezing of the capital investment,
ρ discounted annual depreciation rate.

7.1.1 Specific Cost of Electricity Production in a Power Unit for Supercritical Parameters

By applying Eqs. (7.2), (7.3) and the relations

• fuel cost

$$K_{fuel} = E_{ch,A}\ e_{coal}$$ (7.4)

- net efficiency of electricity production in a power plant

$$\eta_{el} = \frac{E_{el,A}}{E_{ch,A}} = \frac{\int\limits_0^{\tau_A} N_{el} d\tau}{\int\limits_0^{\tau_A} \dot{E}_{ch} d\tau} \cong \frac{N_{el,n}\tau_A}{\dot{E}_{ch,n}\tau_A} \tag{7.5}$$

- maintenance and overhaul cost

$$K_{serv} = \delta_{serv}J \tag{7.6}$$

- investment

$$J = N_{el,n}j \tag{7.7}$$

where:

$E_{ch,A}$	annual use of the chemical energy of the fuel,
$\dot{E}_{ch,n}$	nominal stream of the chemical energy of the fuel,
e_{coal}	specific fuel price (per energy unit),
j	specific investment (per unit of power),
$N_{el,n}$	nominal power of the power plant,
δ_{serv}	annual rate of maintenance and overhaul of equipment relative to the investment,
τ_A	annual operation time of the power plant; the activity period for the new power units is estimated at $\tau_A = 7500$ h/a,

the specific cost (7.1) can be expressed as the total of the specific variable cost and fixed cost associated with the production of electricity in a power plant

$$k_{el} = (1 + x_{A\,var})\frac{e_{coal}}{\eta_{el}} + (1 + x_{A\,fix})\frac{(z\rho + \delta_{serv})j}{\tau_A} \tag{7.8}$$

where:

$x_{A\,fix}$	annual rate of the outstanding fixed cost (payroll with overheads, taxes, insurance, the calculations adopted the value of $x_{A\,fix} = 5$ %,
$x_{A\,var}$	annual rate of non-fuel variable cost (cost of the power plant internal load, cost of supplementing water, cost of non-energy recourses and auxiliary materials, charges for the use of the environment); the calculations adopt ed the value of $x_{A\,var} = 10$ %,
$z\rho + \delta_{serv}$	annual rate of capital maintenance and outstanding fixed cost relative to investment (maintenance and overhaul cost); calculations adopted the value of $z\rho + \delta_{serv} = 16$ %.

By substituting into the Eq. (7.8) the data we gain that the specific cost of electricity production in a power unit for supercritical parameters—Fig. 7.1—with the capacity of $800 \div 900$ MW and the necessary specific investment of $j = 6.5$ million PLN/MW (contracted investment in a power unit with the capacity of $800 \div 900$ MW is equal to 5.5 billion PLN) and the price of coal $e_{coal} = 11.4$ PLN/GJ is equal to

$$k_{el} = 1,1 \frac{11,4\,\text{PLN/GJ}}{0,456} \times \frac{3,6\,\text{GJ}}{\text{MWh}} + 1,05 \frac{0,16 \times 6\,500\,000\,\text{PLN/MW}}{7\,500\,\text{h}}$$

$$\cong 99 + 146 = 245\ [\text{PLN/MWh}].$$

$$(7.9)$$

The variable part of the specific cost of the production of electricity (in which the major share is made up of the cost of fuel) is equal to $k_{el\ var} = 99$ PLN/MWh, while the fixed cost, due to the specific capital investment in a power unit is equal to $k_{el\ fix} = 146$ PLN/MWh.

The specific cost of nuclear fuel in a nuclear power plant is responsible for merely several percent of the specific cost k_{el} of electricity production in them. The fixed cost is equal to $k_{el\ fix} = 345$ PLN/MWh ($j = 18$ million PLN/MW, $\tau_A = 8760$ h/a).

The value of the efficiency $\eta_{el} = 45.6$ % in Eq. (7.9) is imposed by the value of CO_2 emission factor from the power plant. This ratio, which expresses the

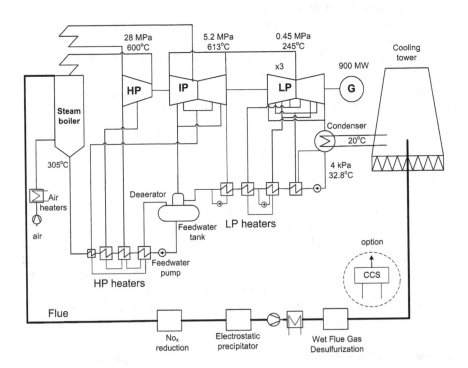

Fig. 7.1 Diagram of a power unit for supercritical parameters

kilograms of carbon dioxide per megawatt-hour of the produced electricity E_{el} from the value of E_{ch} of the chemical energy of the fuel combustion, in accordance with the climate regulations of the EU should not exceed the value of

$$EF_{CO_2} = \frac{E_{ch}\rho_{CO_2}}{E_{el}} = \frac{\rho_{CO_2}}{\eta_{el}} = 750 \left[\frac{kg_{CO_2}}{MWh}\right] \qquad (7.10)$$

where ρ_{CO2} denoted the emission of CO_2 in kilograms per unit of the chemical energy of the fuel combustion in the power plant, and η_{el} denotes the net efficiency of the production of electricity in it. For the black coal, the emission of CO_2 is equal to $\rho_{CO2}^{coal} \cong 95 \ kg_{CO2}/GJ = 342 \ kg_{CO2}/MWh$ (for natural gas, it is $\rho_{CO2}^{gas} \cong 55 \ kg_{CO2}/GJ = 198 \ kg_{CO2}/MWh$). In order to gain the value recommended by the EU, the value of EF cannot exceed $EF_{CO2} = 750 \ kg_{CO2}/MWh$ (however, the values of the emission factor of $EF_{CO2} = 500 \ kg_{CO2}/MWh$ are already mentioned and even lower ones, in the range of $EF_{CO2} = 100 \ kg_{CO2}/MWh$), the net efficiency of the power plant has to be equal to $\eta_{el} = 45.6 \ \%$. Such efficiency is possible to achieve in power plants for supercritical parameters with the value of at least 28 MPa, 600/620 °C. The current value of the emission factor from a 370 MW coal-fired power unit with the net efficiency of $\eta_{el} = 37 \ \%$ is equal to $EF_{CO2} = 924 \ kg_{CO2}/MWh$.

7.1.2 Specific Cost of Electricity Production in a 370 MW Power Unit Adapted to a Gas–Steam System

The specific cost of producing electricity $(k_{el})^{mod}$ in a repowered power plant—Chap. 5, Fig. 5.1—is made up of the weighted mean of the cost $(k_{el})^{ex}$ and $(k_{el})^{\Delta E_{el,A}}$, i.e. cost $(k_{el})^{ex}$ of producing electricity prior to its repowering and the cost $(k_{el})^{\Delta E_{el,A}}$ associated with the increase of its production as a result of the repowering:

$$(k_{el})^{mod} = \frac{(E_{el,A})^{ex}}{(E_{el,A})^{ex} + \Delta E_{el,A}}(k_{el})^{ex} + \frac{\Delta E_{el,A}}{(E_{el,A})^{ex} + \Delta E_{el,A}}(k_{el})^{\Delta E_{el,A}} \qquad (7.11)$$

while

$$(k_{el})^{ex} = \frac{(K_A)^{ex}}{(E_{el,A})^{ex}} \qquad (7.12)$$

$$(k_{el})^{\Delta E_{el,A}} = \frac{\Delta K_A}{\Delta E_{el,A}}. \qquad (7.13)$$

where:

$(E_{el,\ A}\ )^{ex}$ annual net electrical loco output of the power plant prior to its repowering,

$\Delta E_{el,\ A}$ increase of the annual net production of electricity loco power plant after its repowering,

$(K_A)^{ex}$ annual operating cost of the power plant prior to its repowering,

ΔK_A increase of the annual operating cost of the power plant after its repowering.

By substituting (7.12) and (7.13) into (7.11) we obtain that

$$(k_{el})^{\text{mod}} = \frac{(K_A)^{ex} + \Delta K_A}{(E_{el,A})^{ex} + \Delta E_{el,A}} \tag{7.14}$$

The increase in the annual net production of electricity in a repowered power plant is equal to

$$\Delta E_{el,A} = (E_{el,A}^{GT,\,gross} + \Delta E_{el,A}^{ST,\,gross})(1 - \varepsilon_{el}^{\text{mod}}), \tag{7.15}$$

where: $E_{el,A}^{GT,gross}$, $\Delta E_{el,A}^{ST,gross}$ denote, respectively, the annual gross production of electricity in a gas turbogenerator and annual increase of its production in the steam turbogenerator, $\varepsilon_{el}^{\text{mod}}$ power plant parasitic load (the calculations adopted $\varepsilon_{el}^{\text{mod}} = 4\ \%$).

The increase of the annual cost ΔK_A is expressed by the equation

$$\begin{aligned}
\Delta K_A &= (K_A)^{\text{mod}} - (K_A)^{ex} \\
&= (z\rho + \delta_{serv})J^{\text{mod}} + K_{gas}^{GT} + K_{env}^{GT} - \Delta K^{coal} - \Delta K_{r,m,w}^{coal} - \Delta K_{env}^{coal}
\end{aligned} \tag{7.16}$$

where:

J^{mod} turnkey investment on the adaptation of the power unit by its repowering with a gas turbogenerator and heat recovery steam generator,

K_{gas}^{GT} cost of natural gas combustion in the gas turbine,

K_{env}^{GT} charge for the use of environment resulting from the burning of natural gas in the gas turbine,

ΔK^{coal} decrease of the cost of coal purchase,

$\Delta K_{r,m,w}^{coal}$ reduction of the maintenance and overhaul cost, cost of non-energy resources and supplementing water; the calculations can adopt the value of $\Delta K_{r,m,w}^{coal} = 0$,

ΔK_{env}^{coal} reduction of the cost of charges for the use of environment resulting from the reduced volume of coal combustion in the power plant.

The cost of the natural gas combustion in the gas turbine is expressed by the equation

$$K_{gas}^{GT} = E_{ch,A}^{gas} e_g \qquad (7.17)$$

where $E_{ch,A}^{gas}$ denotes annual use of the chemical energy of the gas relative to the capacity of the gas turbogenerator, e_g the specific gas price (per energy unit).

The decrease of the cost of coal purchase in the existing steam boiler is equal to

$$\Delta K^{coal} = \Delta E_{ch,A}^{coal} e_{coal}, \qquad (7.18)$$

where $\Delta E_{ch,A}^{coal}$ denoted the annual reduction in the use of the chemical energy of the coal relative to the capacity of the gas turbogenerator and structure of the heat recovery steam generator, e_{coal}—specific coal price (per energy unit).

The environmental cost K_{env}^{GT} for the gas system and the decrease of the cost ΔK_{env}^{coal} associated with the reduction of the volume of coal combustion in the power plant are relative to the specific charges associated with the use of the environment and can be expressed by the equations

$$K_{env}^{GT} = E_{ch,A}^{gas}\left(\rho_{CO_2}^{gas}p_{CO_2} + \rho_{CO}^{gas}p_{CO} + \rho_{SO_2}^{gas}p_{SO_2} + \rho_{NO_x}^{gas}p_{NO_x}\right), \qquad (7.19)$$

$$\Delta K_{env}^{fuel} = \Delta E_{ch,A}^{coal}\left(\rho_{CO_2}^{coal}p_{CO_2} + \rho_{CO}^{coal}p_{CO} + \rho_{SO_2}^{coal}p_{SO_2} + \rho_{NO_x}^{coal}p_{NO_x} + \rho_{dust}^{coal}p_{dust}\right), \qquad (7.20)$$

where:

$p_{CO_2}, p_{CO}, p_{NO_x}, p_{SO_2}, p_{dust}$	specific charges for CO_2, CO, NO_x, SO_2 and dust emissions, PLN/kg,
$\rho_{CO_2}^{gas}, \rho_{CO}^{gas}, \rho_{NO_x}^{gas}, \rho_{SO_2}^{gas}$	CO_2, CO, NO_x, SO_2 emission per unit of the chemical energy of the gas, kg/GJ,
$\rho_{CO_2}^{coal}, \rho_{CO}^{coal}, \rho_{NO_x}^{coal}, \rho_{SO_2}^{coal}, \rho_{dust}^{coal}$	CO_2, CO, NO_x, SO_2, and dust emission per a specific unit of the chemical energy of the coal, kg/GJ.

The total cost of the environmental charge in the coal-fired system is expressed by the equation

$$\Delta K_{env}^{coal} = \Delta K_{env}^{fuel} + \Delta K_{env}^{non-fuel} \qquad (7.21)$$

The non-fuel cost $\Delta K_{env}^{non-fuel}$ includes the cost of ash and slag utilization, waste storage, use of water and wastewater disposal, purchase and transport of chemicals for water treatment (demineralization and decarbonization), limestone dust and other chemicals for the installation of wet flue gas desulfurization IOS and cost of carbamide for NO_x reduction system.

The computer simulation of the specific cost $(k_{el})^{mod}$ of electricity production in a repowered 370 MW unit applied its mathematical model presented in [2]. By the aid of it, the following values were derived $(E_{el,A})^{ex}$, $E_{el,A}^{GT,gross}$, $\Delta E_{el,A}^{ST,gross}$, $E_{ch,A}^{gas}$ and $\Delta E_{ch,A}^{coal}$. All these variables, except for $(E_{el,A})^{ex}$ are the function of the capacity of the gas turbogenerator and the structure of the heat recovery steam generator. The calculations of the specific cost of the production of electricity applied Eq. (7.14) in the form

$$(k_{el})^{mod} = \frac{(E_{el,A})^{ex}(k_{el})^{ex} + \Delta K_A}{(E_{el,A})^{ex} + \Delta E_{el,A}} \tag{7.22}$$

while the cost $(k_{el})^{ex}$ prior to the repowering of the 370 MW unit with the net efficiency of $\eta_{el} = 37\%$ is assumed to be at 170 PLN/MWh, and its variable part is equal to $(k_{el\ var})^{ex} = 125$ PLN/MWh, while the constant part is $(k_{el\ fix})^{ex} = 45$ PLN/MWh.

The calculations of the specific cost $(k_{el})^{mod}$ have adopted the following values of input data:

- estimated turnkey investment in the adaptation of the power unit by its repowering by a gas turbogenerator with the capacity of 350 MW and a triple-pressure heat recovery steam generator: $J^{mod} = 570$ million PLN,
- relative coefficient of parasitic load of the repowered unit $\varepsilon_{el}^{mod} = 4\%$
- specific coal price $e_{coal} = 11,4$ PLN/GJ
- specific gas price $e_g = 28$ PLN/GJ
- annual rate of depreciation, maintenance and overhaul $z\rho + \delta_{serv} = 16\%$
- specific emission charges: $p_{CO2} = 0.25$ PLN/Mg, $p_{CO} = 0.11$ PLN/kg, $p_{NOx} = 0.46$ PLN/kg, $p_{SO2} = 0.46$ PLN/kg.

The overall capacity of the power unit after its repowering is equal to 800 MW. As a result of the modernization the capacity of the steam turbogenerator increases as well as a consequence of the increase of the capacity of its low-pressure stage. The capacities of the intermediate- and high-pressure stages alter inconsiderably. Hence, new low-pressure stage and condenser with a higher throughput as well as an electrical generator with a higher capacity are needed, which is already included in the investment necessary for the adaptation of the power unit.

The calculated value $(k_{el})^{mod}$ for the power unit repowered by the gas turbogenerator with the capacity of 350 MW and a triple-pressure heat recovery steam generator— Chap. 2, Fig. 2.2—is equal to 189.7 PLN/GJ—Figs. 7.2, 7.3.

7.1.3 Analysis of Sensitivity

After calculation of the specific cost of electricity production $(k_{el})^{mod}$ it is necessary to conduct an analysis of its sensitivity in the function of the parameters that

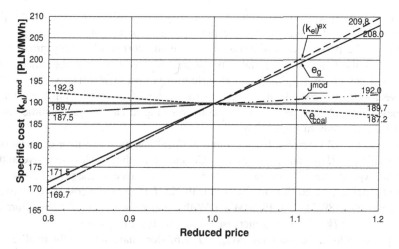

Fig. 7.2 Influence of coal and gas prices, cost of producing electricity in power unit prior to its repowering and capital expenditure on the value of specific cost of electricity production in a repowered power unit

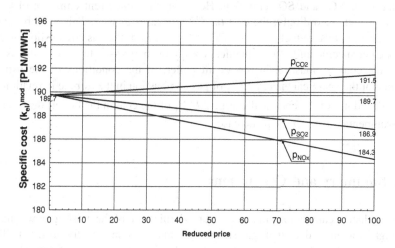

Fig. 7.3 Influence of specific emission charges on the value of specific cost of electricity production in a repowered power unit

affect them. The analysis of sensitivity offers a wider perspective with regard to the profitability of an investment and enables an investor to assess its security as well as offers grounds for conducting price policy in the conditions of a competitive market. It presents the range of primary fuel prices which will secure the profitability of an undertaking and the scope for reducing the price of a product that will ensure that they will not go of business. This level is determined as zero profits level, i.e. corresponds to zero value of $(e_{el})^{mod}$—$(k_{el})^{mod}$ where $(e_{el})^{mod}$ denotes the specific sale price of electricity from the repowered power unit.

Figure 7.2 present the variations of the specific cost of electricity production $(k_{el})^{mod}$ in the function of capital expenditure J^{mod} as well as in the function of coal and gas prices e_{coal}, e_{gas} and electric energy cost $(k_{el})^{ex}$. The values of the above taken into consideration vary in the range of $\pm$ 20 % from their basic values—Fig. 7.2. The reduced prices corresponding to basic prices assume value of 1 on the X axis in Figs. 7.2 and 7.3.

From Fig. 7.2 it results that the value of $(k_{el})^{mod}$ is most sensitive to variations of specific cost $(k_{el})^{ex}$ and gas price e_{gas}. Concurrently, they are less sensitive to the capital expenditure J^{mod} and coal price e_{coal}. For instance, if the price of gas e_{gas} were to increase by 20 % from 28 to 33.6 PLN/GJ, the specific cost $(k_{el})^{mod}$ would increase from 189.7 to 208 PLN/MWh. With the increase of capital expenditure by 20 % the cost $(k_{el})^{mod}$ would increase merely to 192 PLN/MWh.

Specific allowances on CO_2, NO_x, SO_2 emissions have an equally small effect on the cost $(k_{el})^{mod}$—Fig. 7.3. Even a 100th increase of these charge does not lead to a considerable increase of this cost. This is so since the existing power plants already have installations for flue gas desulfurization and denitrogenation while the gas turbine burns an ecological fuel, i.e. natural gas, which leads to a slight reduction in NO_x and SO_2 emission. However, the concurrent emission of CO_2 increases and, accordingly, the cost $(k_{el})^{mod}$ increases as well. This is so because the almost doubled reduction of the CO_2 emission from the gas combustion in gas turbine per specific unit of the chemical energy in comparison to the CO_2 emission from the coal combustion in the steam boiler, in the conditions of the tripled volume of the chemical energy $E_{ch,A}^{gas}$ of gas combustion in the turbine in relation to the reduced use of the chemical energy $\Delta E_{ch,A}^{coal}$ of the coal in the boiler, the overall environmental cost CO_2 will increase.

7.2 Summary and Conclusions

The conducted calculations indicate that modernization of the existing power units to high-efficiency dual-fuel gas-steam systems is economically justified. The specific cost of producing electricity in them despite the high price of natural gas is considerably smaller than the cost of construction of new, expensive power units for supercritical parameters. What is more, carbon dioxide emission factor from a dual-fuel system is significantly smaller from supercritical power units and is equal to around $EF_{CO2} = 500$ kg$_{CO2}$/MWh.

References

1. Bartnik R (2009) Combined cycle power plants. Thermal and economic effectiveness. WNT, Warszawa
2. Bartnik R, Buryn Z (2011) Conversion of coal-fired power plants to cogeneration and combined-cycle. Thermal and economic effectiveness. Springer, London
3. Bartnik R (2008) Technical and economic efficiency account in utility power engineering. Oficyna Wydawnicza Politechniki Opolskiej, Opole

Chapter 8
Summary and Final Conclusions

Abstract This chapter presents the most important general remarks arising out of the discussion of the all issues presented in the monograph.

Keywords General remarks · Coal-fired power unit · Repowering · Gas turbine · Heat recovery steam generator · Dual-fuel combined-cycle

The conversion of a 370 MW power unit into a dual-fuel system enables an increase of the efficiency of production of electricity by around 10 %. The increase of the efficiency and the combustion of ecological natural gas beside the coal also ensures the reduction of CO_2 emission per specific unit of the produced electricity by as much as 50 %.

The conversion of a 370 MW power unit into a dual-fuel system enables the doubling of its overall capacity. A further remark is associated with the fact the conversion can take as little as several months.

The thermodynamic criterion for the selection of a gas turbogenerator for the repowered unit and the structure of the heat recovery steam generator (number of pressure stages in it) and the installed heating surfaces and their arrangement is the one associated with the maximization of the efficiency of the production of electricity. From the multiple alternatives of computer calculation undertaken for the purposes of this study it stems that the optimum capacity of the gas turbogenerator is mainly relative to the type of the heat recovery steam generator (number of pressure stages) and to a lower degree to the temperature of the exhaust gases from it. The higher the number of the pressure stages, the higher the optimal capacity of the gas turbogenerator, since the efficiency of the production of electricity in the repowered unit is higher.

In comparison to the energy efficiency of the power unit, the more superior criterion for the repowering of a unit to a dual-fuel gas-steam system should be the one associated with the economic profitability of the undertaking and maximization of the profit from the operation of a repowered power plant. This one should

R. Bartnik, *The Modernization Potential of Gas Turbines in the Coal-Fired Power Industry*, SpringerBriefs in Applied Sciences and Technology, DOI: 10.1007/978-1-4471-4860-9_8, © The Author(s) 2013

ultimately decide about the capacity of the installed gas turbine and structure of the heat recovery steam generator. One has to bear in mind that the dual-fuel gas-steam technology offers an opportunity to modernize the existing coal-fired power plants with the smallest investment. Such modernization needs four times lower expenditure in comparison to the construction of a new power unit operating under supercritical parameters calculated per specific unit of electrical capacity.

Index

R. Bartnik, *The Modernization Potential of Gas Turbines in the Coal-Fired*
Power Industry, SpringerBriefs in Applied Sciences and Technology,
DOI: 10.1007/978-1-4471-4860-9, © The Author(s) 2013